Geometrical Enhancements Based on Data-Driven Design Decisions

The text presents various design and modelling solutions for effective decision-making that are grounded on the basics of data analytics. It further discusses important topics such as sustainable design and data-driven design synthesis, product analytics and its role in sustainable development, and descriptive models for the geometrical evaluation of product features.

This book:

- Familiarizes the readers with the numerous decision protocols that can be utilized for deducing geometrical qualities using data-driven design analysis.
- Offers a more comprehensive approach to multidisciplinary engineering applications using data-driven modelling and analysis.
- Discusses multi-objective optimization for product design and development through data-driven decision-making.
- Explains information processing and analysis through data visualization and data mining.
- Covers artificial intelligence and machine learning-based decision models and their effective utilization across interdisciplinary applications.

It is primarily written for senior undergraduate, graduate students, and academic researchers in fields such as industrial engineering, production engineering, manufacturing engineering, mechanical engineering, and aerospace engineering.

Geometrical Enhancements Based on Data-Driven Design Decisions

Edited by
Sourabh Mandol, Mohit Hemanth Kumar,
Akarsh Verma and V. K. Singh

CRC Press
Taylor & Francis Group
Boca Raton London New York

CRC Press is an imprint of the
Taylor & Francis Group, an **informa** business

First edition published 2025
by CRC Press
2385 NW Executive Center Drive, Suite 320, Boca Raton FL 33431

and by CRC Press
4 Park Square, Milton Park, Abingdon, Oxon, OX14 4RN

CRC Press is an imprint of Taylor & Francis Group, LLC

ISBN: 978-1-032-65500-0 (hbk)
ISBN: 978-1-041-02825-3 (pbk)
ISBN: 978-1-003-62110-2 (ebk)

DOI: 10.1201/9781003621102

Typeset in Sabon
by SPi Technologies India Pvt Ltd (Straive)

Contents

About the Editors

Sourabh Mandol is a skilled data analyst and researcher with a PhD in Product Design and Development from IIT Kharagpur, India. Specializing in data-driven modelling and machine learning applications in engineering, Dr. Mandol has extensive expertise in applying advanced analytical techniques, including statistical analysis, soft computing, and multidisciplinary design optimization. His research focuses on developing innovative methodologies for complex engineering systems, particularly in the areas of eco-design, sustainability, and hybrid electric vehicle optimization. Currently serving as an assistant professor, Dr. Mandol brings his industry-relevant expertise to the classroom, teaching courses in data-driven decision-making, analytics, and product design. He has a proven track record of implementing machine learning algorithms for quality control and process optimization in manufacturing, as well as developing digital twin concepts for engineering applications. Dr. Mandol's work has been published in reputable journals and presented at international conferences, showcasing his commitment to driving innovation through the integration of data science and engineering principles.

Mohit Hemanth Kumar is an accomplished academic currently serving as Assistant Professor and Associate Director of the AU IPR Cell at Alliance University, Bengaluru. With 5.5 years of teaching and research experience, he earned his PhD from the National Institute of Technology, Tiruchirappalli, where he specialized in the development and investigation of nanofiller-reinforced hybrid polymer composites. His research interests span several critical areas in mechanical engineering and materials science, focusing on inorganic filler-reinforced hybrid polymer composites, nanocomposites, waste-to-innovative products, industrial product design and development, and manufacturing technology. As an educator, he has developed and taught courses in mechanical engineering fundamentals, product development and design, additive manufacturing, smart manufacturing, Digital Manufacturing and Rapid Prototyping, Big Data Analytics for Manufacturing, Smart Materials and Structure, and materials technology. Dr. Kumar's research contributions have significantly impacted his field, garnering over 1,500 citations and achieving an h-index of 19. His innovative work has led to 24 granted patents and 25 granted industrial designs, with 5 more patents awaiting approval. He has published 43 articles in reputed international journals, authored 34 book chapters, and served as editor for 6 books. As a registered Patent Agent (IN/PA 5341), Dr. Kumar actively contributes to intellectual property development and protection. He maintains strong professional affiliations as a Life Member of several prestigious organizations, including the Indian Society of Technical Education, Institute of Engineer's India, Society for Failure Analysis, International Association of Engineers, and the Powder Metallurgy Association of India. Through his international research collaborations and contributions to curriculum development, Dr. Kumar continues to advance both the academic and practical aspects of mechanical engineering and materials science.

Akarsh Verma is currently an assistant professor at the University of Petroleum and Energy Studies (UPES), India. He obtained a PhD in Mechanical Engineering from the Indian Institute of Technology (IIT) Roorkee, India, and did his postdoctoral research work at the Brigham Young University (USA) and was a JSPS International Research Fellow at the Osaka University (Japan). He serves the research community by acting as a reviewer for various international journals and has reviewed several research articles. In addition, he has published various research articles in high-quality international peer-reviewed journals, many books and book chapters, and also presented many research papers at various international/national conferences. His areas of interest are computational mechanics, computational chemistry, quantum mechanics, molecular dynamics, and nanocomposites. He has various international fellowships/awards (Japan Society for the Promotion of Science (JSPS) Postdoctoral Fellowship, Japan; J. William Fulbright Doctoral Research Fellowship, USA; Early Career Training Grant award for Early Career Researchers from the Communications Materials (*Nature* Journal); Researcher Development Grant from the Royal Society of Chemistry (RSC), UK; Excellence in Doctoral Research Award for the Best PhD thesis research work at IIT Roorkee, India; Doctoral Research Scholarship by the Department of Science and Technology (DST), Government of India and Ministry of Human Resource Development (MHRD), Government of India; PG Scholarship by the All India Council for Technical Education (AICTE), India; Young Researcher Award ICMR2022, Japan) to his credit. He has also been recognized by Stanford University's list of the world's Top 2% of the Most-Cited Scientists in Single Year Citation Impact (2023, 2022, 2021, and 2020).

V. K. Singh is currently a professor in the Department of Mechanical Engineering at Govind Ballabh Pant University of Agriculture and Technology, Pantnagar, India. Prof. Singh has over 20 years of teaching and research experience. He has contributed to over 100 papers in various international and national journals and conferences. He has also overseen several research projects and also guided a number of MTech and PhD theses. His current research interests are in the areas of polymers, composite materials, and fracture mechanics. He is also a member of several engineering societies and serves the research community by reviewing research papers. He has also organized over 10 international and national conferences. He has also served in administrative-level positions at the university and state levels.

Contributors

Pallab Banerjee
Amity University
Jharkhand, Ranchi, India

Yadavalli Basavaraj
Department of Mechanical Engineering
Ballari Institute of Technology & Management
Ballari, Karnataka, India

Debraj Bhattacharjee
Flame University
Maharashtra, India

Sourajit Behera
Kalinga Institute of Industrial Technology
Bhubaneshwar, India

Santosh Kumar Biswal
Department of Journalism and Mass Communication,
Rama Devi Women's University
Odisha, India

Uttam Chakraborty
School of Management
Presidency University
Karnataka, India

R. Dhaya
Department of Electrical and Communication Engineering,
PNG University of Technology,
Lae, Papua New Guinea

Pankaj Dixit
School of Management Studies
National Forensic Sciences University, Gandhinagar
Gujrat, India

R. Adline Freeda
Dept. of Information Technology
KCG College of Technology, Chennai, India

Kumar Gaurav
Amity University
Jharkhand, Ranchi, India

Hemant Gupta
ICFAI Business School Bengaluru,
Karnataka, India

Juhi Gupta
Amity Institute of Environmental Sciences
Amity University
Noida, India

Naman Jain
Department of Mechanical Engineering,
ABES Engineering College, Ghaziabad,
Uttar Pradesh, India

Naman Joshi
Department of Mechanical Engineering
ABES Engineering College, Ghaziabad
Uttar Pradesh, India

Aditya Kataria
Department of Mechanical Engineering
University of Minnesota-Twin Cities
Minneapolis, United States of America

N. Keerthi Kumar
Department of Mechanical Engineering
BMS Institute of Technology and Management
Bangalore, Karnataka, India

R. Kanthavel
Department of Electrical and Communication Engineering,
PNG University of Technology,
Lae, Papua New Guinea

K. Lokesha
Department of Mechanical Engineering,
M S Ramaiah Institute of Technology,
Bangalore, India

C. M. Manasa
Manipal Institute of Technology Bengaluru
Manipal Academy of Higher Education
Manipal, India

Rushali Mishra
Amity Institute of Environmental Sciences
Amity University
Noida, India

Dipra Mitra
Amity University
Jharkhand, Ranchi, India

Himanshu Parashar
Department of Mechanical Engineering
University of Petroleum and Energy Studies
Dehradun, India

Bhaveshkumar J. Parmar
School of Management Studies
National Forensic Sciences University, Gandhinagar
Gujrat, India

Ajit Kumar Pasayat
Kalinga Institute of Industrial Technology
Bhubaneshwar, India

Jayshree Patnaik
Indian Institute of Technology Hyderabad
Telangana, India

B. K. Pavan Kumar
Department of Mechanical Engineering
Ballari Institute of Technology & Management
Ballari, Karnataka, India

Malesh Raghuvanshi
Department of Mechanical Engineering,
ABES Engineering College, Ghaziabad,
Uttar Pradesh, India

Sudeshna Sani
Woxsen University, Hyderabad,
Telangana, India

Anurag Sharma
Department of Mechanical Engineering,
ABES Engineering College, Ghaziabad,
Uttar Pradesh, India

Pratishtha Sharma
Department of Mechanical Engineering
Indian Institute of Technology, Delhi
Delhi, India

Prashant Shukla
School of Advanced Engineering,
University of Petroleum and Energy Studies
Dehradun, India

Prashant Upadhyay
Kalinga Institute of Industrial Technology
Bhubaneshwar, India

V. Venkataramana
Department of Mechanical Engineering
Ballari Institute of Technology & Management
Ballari, Karnataka, India

Akarsh Verma
Department of Mechanical Engineering
University of Petroleum and Energy Studies
Dehradun, India

Chapter 1

Introduction to open visualization tool (OVITO)

Aditya Kataria, Himanshu Parashar and Akarsh Verma

1.1 INTRODUCTION

OVITO is a data analysis and visualization tool [1, 2], with regards to molecular and other particle-based simulation models, widely used in computational materials science, chemistry, and physics.

The program is established, circulated, and supported by OVITO GmbH, a Germany-based spin-off company founded by Dr. Alexander Stukowski, the creator of the novel program OVITO. The software was first made available to the public in 2009; however, the software was developed a year prior as a component of a PhD project for the Materials Science Department at Technische Universität Darmstadt. Up until 2020, Dr. Stukowski worked part-time at TU Darmstadt while continuing his development [3].

1.2 BASIC USAGE

1.2.1 Data import

Use the corresponding toolbar button or choose "Load File" from the menu to load a simulation file into a computer system. OVITO can read compressed text files with a ".gz" extension directly and is capable of automatically recognizing various file types. The imported dataset will appear as a visual object in the pipeline editor's "Data source" section and the 3D viewports.

The pipeline editor has an external file panel just underneath it. It features tool buttons to change the file used as the source of the data pipeline and, in case an imported file was updated outside of OVITO, reload it. Changing this file is useful when setting up a pipeline that contains complex data analysis and for using it again on a new simulation dataset.

1.2.1.1 Command line usage

When launched from the terminal, OVITO can load the files straight away. This works for local files and those that are remote. Providing various paths makes it simple to import numerous files at once. They will be combined

DOI: 10.1201/9781003621102-1

into an animatable trajectory if they are all in the same data format. If the formats differ, OVITO will identify them as a pair of topology/trajectory files. If not, they will be added to the scene as stand-alone objects.

1.2.1.2 Simulation trajectories

OVITO can load the simulation trajectories that are formed from a series of snapshots. The software supports many scenarios:

CASE [1]: A SERIES OF SEPARATE FILES

By default, OVITO looks in the directory to look for new simulation files to see if they are a part of a series of files, that are serialized numerically with similar names. This is accomplished by substituting the wildcard character "*" for the last alphanumeric character arrangement in the filename chosen. This creates an examination pattern that can be used for future work to find further files in the same directory that are part of the trajectory sequence. For instance, OVITO will create the search pattern anim1c *.dump when we open a file called "anim1c 200.dump" to locate more snapshots, such as anim1c 0.dump, anim1c 400.dump, anim1c 700.dump, and so on.

CASE [2]: ONE FILE CONSISTING OF ALL TRAJECTORY FRAMES

OVITO makes a totalized animation sequence when it recognizes that an imported file has several simulation frames. Typically, OVITO only keeps one frame's worth of data in memory at a time. Only, when necessary, like when the animation is played back or the time slider is adjusted, are the subsequent frames put into memory.

CASE [3]: PAIR OF TRAJECTORY AND TOPOLOGY FILES

Certain MD systems of simulations require various document files depending on the molecular structure's trajectory along with its topology. The topology file offers static descriptions of atoms, bonds, and other components, while the trajectory file includes calculated trajectories and other data, generally time-dependent, produced during the dynamic simulation. The topology file, which is a LAMMPS data file, has been loaded via a load trajectory modifier in the data pipeline. This change can unify both the dynamic trajectory data and static topology and convert it into one animated dataset.

OVITO automatically detects and differentiates the topology file and the trajectory file. For more format details, visit Data import—OVITO User Manual 3.10.6 documentation.

1.2.1.3 Visualizing multiple datasets

OVITO may be used to create three-dimensional scenes by combining many objects or information. It is possible to import and see many datasets in one picture. The same dataset may also be seen in various ways, such as side by side or by superimposing several display options. An imported dataset is dynamically replicated using OVITO's "branched data pipelines" functionality, treating each clone differently.

Using the "Load File" command several times to import each dataset into the current scene is the easiest approach to integrating multiple datasets in a single image. When another dataset is imported, OVITO asks if the user wants to replace the dataset that was previously imported. To add the following datasets as additional objects to the scene, use the "Add to scene" option. All of the datasets and other objects in the current scene are listed in the top toolbar under the "pipeline selection" widget. Every imported dataset has its data pipeline attached to it. Consequently, each of them can be subjected to various alterations.

1.2.1.4 Positioning objects

The same standard location in the scene's global coordinate system is used when OVITO assigns all imported datasets. This may lead to an unpleasant situation when many datasets are initially piled in the same spatial location. This may be changed by manipulating individual elements and rearranging the scene.

Activating the "Translate" input mode allows the movement of objects about in the viewports using the mouse. Alternatively, the necessary XYZ location for the current item using the input boxes in the status bar may also be entered. To transmit a subset of the dataset, such as a collection of chosen particles, rather than the complete dataset, use the "Affine transformation" modification in the "data pipeline" [4, 5].

1.2.2 Viewport windows

OVITO's viewport windows show many perspectives of the scene in three dimensions. One viewport that fills the main window as well as the standard two-by-two grid arrangement are available. The subtitle in each viewport's upper left corner and the axis tripod in the lower left corner both show the orientation of the virtual camera.

1.2.2.1 Camera navigation

Use the mouse to move or tilt a viewport's virtual camera:

1. Using the left-click along with drag allows the user to rotate the camera around the current orbit center. The orbit center is by default in the center of the model.
2. Using a right-click along with a drag gesture with the mouse to transfer the camera parallel to the projection plane. Alternatively, it can be accomplished by using the "Shift + left" or center mouse buttons.
3. Using the mouse wheel for zooming in and out.
4. Double-clicking on any item will relocate the orbit center to the area beneath the mouse cursor. Now that new point—which is represented by a three-dimensional cross—will be centered by the camera.
5. To return the orbit center to the center of the simulation model, double-click on a viewport's empty area.

1.2.2.2 Viewport toolbar

The buttons on the viewport toolbar, which is positioned underneath the viewports, can be used to specifically enable different navigation input choices. Additionally, there are two more crucial techniques, as follows:

The button "Zoom Scene Extents" will automatically adjust the active viewport's virtual camera to make every object in the scene visible. Holding down the "Ctrl" key allows the user to do it for every viewport at once.

The active viewport may be made to occupy the whole main window by clicking the button titled "Maximize Active Viewport." Pressing the button again will bring the viewport layout back to its original 2-by-2 configuration.

1.2.2.3 Viewport menu

Click the caption label in a viewport's top left corner to bring up the viewport menu, as seen in the screenshot.

Using the "View Type" option, one may alternate between perspective and parallel (orthogonal) projection types and standard viewing orientations. The "Adjust View" allows the user to adjust the viewport's camera's location and orientation via numeric input fields.

A virtual frame representing the area that will be shown in produced pictures is displayed in the viewport when the "Preview Mode" option is selected. The frame's aspect ratio corresponds to the picture size that is presently specified in the Render options box. Scene elements and any viewport layers in preview mode will appear within the window named interactive viewport precisely as those would in the concluding output picture.

In the three-dimensional scene, construct a camera object by using the "Create Camera" command. The viewport is modified when the camera object is shifted because the viewport and this object are connected. Because of this, it is possible to animate the camera [6].

1.2.3 Data model

An overview of the OVITO data model, which is used to describe particle-based datasets such as molecular structures, is given in this section. Working with the program's data analysis and visualization tools, which use particle simulation data, requires an understanding of the basic ideas behind OVITO's data model.

1.2.3.1 Particle properties

Each particle has numerical values connected to it that are known as particle characteristics. They are essential to the depiction of molecules and other structures in OVITO's data model. Frequent particle properties include velocity, chemical composition, and particle location. Particles can have any number of extra properties assigned to them by the user, either directly or as a consequence of calculations made by the software.

This fundamental property notion is applied not just to particles but also to other kinds of data components. Interatomic bonding may be related to binding characteristics such as bond type and color, for instance. Although the following discussion focuses on particles and their characteristics, the same concepts hold true for other types of data components.

1.2.3.2 What's a particle property?

An even data array that contains a numerical value for every particle present in the simulation system is referred to as a particle property in OVITO. The data type of each value is the same and can be either vectorial or scalar. The number of particles in the system is always equal to the length of the property array. Each characteristic, such as "Position" or "Potential Energy," has a unique name. Although OVITO provides an integrated list of frequently used property names, one may declare additional properties with user-defined names. Each property name has a specific meaning for the application and a particular data structure.

The unique quality of the "Position" particle feature is its constant presence. Particles cannot exist without spatial coordinates. "Color," "Radius," and "Selection" are extra standard attributes. OVITO displays the particles differently depending on the standard attributes. Changing the settings for these parameters will change how the particles appear visually.

Values for per-particle properties in OVITO can be scalar, vector, or tensor data types, among other dimensionalities. The "Position" is a vector property that has three coordinates—"Position.X," "Position.Y," and "Position.Z"—per particle.

1.2.3.3 Inspecting properties

Typically, one uses the imported simulation file to populate common characteristics such as "Position," "Particle Type," and "Velocity." The extended XYZ format and "LAMMPS dump" are two examples of file formats that support storing an endless number of additional columns containing data. Access the Data Inspector panel in OVITO to view the properties that are currently associated with the particles. Another way to see a particle's property values in the status bar of OVITO is to just move the mouse pointer over it within the viewports.

1.2.3.4 Assigning property values

OVITO offers several capabilities to modify the properties of particles. One may apply a consistent color to all of the currently selected particles by using the "Assign color" adjustment. To achieve this, it sets the particles' "Color" property to the specified RGB value. The subgroup of now chosen particles is found by the "Selection" particle characteristic: The current range set contains particles having a non-zero "Selection" attribute, whereas particles with a "Selection=0" attribute are not.

The "Coordination analysis" modifier determines how many neighbors each particle has inside a certain cutoff array. The outcomes are then saved in a new particle feature named "Coordination." Then, depending on the values of this attribute, particles can be colored according to their coordination number or selected based on their coordination number within a given range. OVITO has several output formats available [7].

1.2.4 Concept of pipeline

The "data pipeline" idea is the foundation of OVITO's operation. It indicates that OVITO is being used to specify a sequence of actions that are applied to the imported simulation dataset one at a time.

OVITO offers a wide range of modifier functions, each of which carries out a particular task like filtering, coloring, analysis, or particle structure selection. By a combination of these essential structure blocks and then setting these up in a pipeline order, it is possible to see or analyze the simulation dataset provided precisely on the user's needs.

One main advantage of this method is that you may modify, remove, or update the applied processes later on if needed, thanks to the data pipeline. Your modifications to the dataset won't stay in effect forever. Any time, more operations can be added, or certain actions can be changed after the fact or eliminated from the pipeline. Every time a change is made to the pipeline, OVITO recomputes the pipeline's output and updates the display to reflect these modifications.

Contemporary picture editing and 3D modeling software, for instance, employ comparable "non-destructive" workflow strategies. With OVITO, one has a great deal of flexibility in creating and experimenting with unique settings for data analysis and visualization, thanks to this technique [8, 9].

1.2.4.1 Pipeline editor

It shows the existing data pipeline and allows one to add new modifiers, reorder them, and change their settings. There are three sections to the editor:

- Visual elements, which are the graphical essentials that the data pipeline generates.
- Modifications, which specify the pipeline's present sequence of modifiers.
- Data source presents data for the pipeline's input, which is typically an external data file.

A drop-down list categorized as "Add modification..." at the upper of the pipeline editor contains these possible modifiers for data. By choosing a modifier from that box, one may add it to the pipeline above the pipeline item that the user presently has selected. With the foremost modifier at the bottom, working on the input data first, the data pipeline is arranged in a bottom-up fashion. By unchecking the box next to a modifier in the pipeline, one can temporarily deactivate it. Use the "Delete Modifier" button located on the right-hand toolbar to eliminate a modifier from the pipeline permanently. Use drag-and-drop actions or the up/down arrow controls in the toolbar to modify the modifiers inside the pipeline if a sequence in which activities are completed is crucial.

1.2.4.2 Visual elements

A dataset including many distinct data pieces, such as particle positions, bond topology, and simulation cell shape, is frequently sent into the pipeline by the data source. Additional information, such as displacement vectors produced from particle trajectories, may be dynamically contributed by pipeline modifiers. The viewports need to provide a visual representation from all of these various types of data leaving the pipeline. The visual components have power over this. The pipeline system automatically creates the visual elements, which are all located in the pipeline editor's top section. The settings of each element are modifiable in the corresponding parameter panel, which governs how the data visualization looks. Visibility of individual items in viewports and generated images

The visual components are in charge of this. The pipeline system automatically generates all of the visual elements, which are all shown at the top of the pipeline editor. Each element's settings, which are modifiable in the corresponding parameter panel, determine how the data visualization looks.

1.2.4.3 Transient pipeline states and particle selection

Few modifiers, such as the Delete selected modification, are limited to the presently selected subset of particles or bonds. The value of an element's "Selection" property determines which items are currently selected. The application has several modifiers that can be used to pick particles depending on various parameters. It's critical to realize that selection occurs dynamically as part of the data pipeline's processing steps: A first modifier selects a subset of these particles, after which another second modifier functions on that particular group. This option will remain the same within the pipeline until it is replaced by another modifier later down the pipeline.

With this approach, the user can sequentially apply diverse processes to different clusters of particles.

It's crucial to remember that data visualization is always restricted to the pipeline's outputs at the end. That is, the user will only be able to see the particle's last choice state in the viewports that are highlighted with red. The subsequent modifiers in the pipeline editor can be temporarily deactivated to analyze a transient condition at any transitional point in the pipeline [9].

1.2.5 Rendering

The picture rendering for the active viewport is initiated by clicking on the "Render active viewport" button. The generated image will be shown by OVITO in a new window, where it may be copied to the clipboard or saved to a disk. The "Render settings" window allows the user to choose the resolution and backdrop color of the final picture. Once rendering is complete, the results are manually saving the image or the user can provide a filename in the Render options box so that the resulting image or video is saved to disk automatically [10].

1.2.6 Miscellaneous

1.2.6.1 Saving session state

A ".ovito" state file, which includes the data pipeline, viewports, render settings, and other information, may be used to store the current application session (primary menu: File Save Session State). The current visualization setup may now be saved for later use. For instance, a state file can serve as a template for displaying several simulations that share a similar camera configuration and data analysis pipeline.

1.2.6.2 Spinner controls

Throughout the program, a spinner widget is a user interface element that may be used to alter numerical values. To change the parameter value, use the spinner widget as indicated below:

1. Click the up arrow once to raise the value, and click the down arrow once to decrease it step-by-step.
2. As an alternative, the user can continuously update the value by clicking and holding the mouse button. The user may move the cursor up or down to change the value of the parameter.

1.2.6.3 Data inspector

In OVITO's main window, a panel located just beneath the viewing area is called the Data Inspector. Several tabs inside the data inspector show dissimilar parts of the present dataset, including all particle property values. The tool may also be used to calculate the distance between two particles.

1.2.6.4 Modifier templates

The user can discover that they frequently utilize the same modifiers when working with OVITO. Certain modifiers are often used with one another to accomplish certain filtering, analysis, or visualization tasks. One may construct modification templates using OVITO to streamline tasks and eliminate needless labor.

1.2.6.5 Python scripting

The user may automate analysis and visualization tasks with OVITO's scripting interface. This might be helpful when handling a big number of input files in batch mode. Most of the program's functionality, including data file input and output, modifiers, and picture and movie rendering, are accessible through the scripting interface. Programming in Python is used to write OVITO programs.

Apart from automating tasks, the user can also enhance OVITO using the scripting interface [11].

1.2.7 Exporting data

The output of the active data pipeline is saved to a file using OVITO's "File ‿ Export File" function. Depending on the output format selected, various parts of the dataset are exported, including bonds, particles and their respective characteristics, and other calculated quantities. Additionally, the user may select whether to export the current frame or a range of frames, and whether to save the datasets to a single output file or a series of files, one for each frame. The user will be prompted by OVITO to provide the destination filename. For text-based file formats, the output file will automatically be compressed if the ".gz" suffix is included.

1.2.7.1 Global attribute values

Several of OVITO's analytical functions, such as the total number of atoms of a certain class or the estimated surface area of a solid, provide scalar output values. On the Attributes page of the data inspector panel, the user may view a table of these global attributes that are associated with the current dataset. Because attributes may have values that vary over time, the pipeline system must dynamically determine each animation frame's attribute value.

The user can export the time series produced by one or more of these global characteristics to a text file by using OVITO's "file export" feature. Make sure to pick the "Table of Values" export format in the file selection popup. Plotting the values of the chosen characteristics against time, this output format creates a tabular text file. The exported data may be plotted using any external charting tool [12].

1.2.7.2 Supported output file formats

To view the supported formats, visit Exporting data—OVITO User Manual 3.10.6 documentation.

1.3 ADVANCED TOPICS

1.3.1 Semi-transparent particles

The "Transparency" particle feature can be used to make particles semi-transparent. A particle with a Transparency value of 0.0 appears entirely opaque, but any value between 0.0 and 1.0 appears transparent. The Compute property modifier is the simplest way to set the Transparency property. To apply the needed transparency value to all particles, simply input it in the expression box, and the Compute property update will take care of the rest.

Sometimes, semi-transparency is needed in only some of the particles. Choose the particles first, then apply the Compute property modifier to achieve this. Turn on its "Compute just for chosen particles" option to restrict the Transparency value's assignment to the currently selected particles alone. Unselected particles will maintain their transparency configurations. Using the Compute property change and the simple arithmetic formula "ReducedPosition.X," the Transparency property was set. Rather than giving every particle the same transparency value, each particle's value is calculated separately by dividing its X location by the simulation cell size, which varies from 0 to 1 based on which side of the simulation box the user is on [13].

1.3.2 Adding a scale bar to a viewport

Even though OVITO cannot add a length scale to an image, an OVITO Python script may be used to paint a custom scale bar on top of a viewport or output picture [14].

1.3.3 Non-spherical particle shapes

Using OVITO, particles with ellipsoid and box forms may be seen. There are also visible particles that have spherocylindrical and cylindrical (capsule) shapes. The "Radius particle" feature regulates the magnitude and form of typical circular and cubic particles, whereas the "Aspherical shape" characteristic controls the magnitude and form of the non-spherical and non-cubic particles. OVITO shows non-spherical particles when this parameter has to be well-defined and the required form needs to be chosen in the Particle visual element's settings.

The radius of ellipsoidal or box particles alongside the three main axes is indicated by the X, Y, and Z components of the "Aspherical Shape" particle characteristic. Particles that have zero values for any of the three components of the Aspherical Shape attribute will be shown as standard spheres (or cubes). Specific to this case, their sizes are determined by the "Radius" attribute, which acts by undefining the "Aspherical Shape" property.

The size of spherocylindrical and cylindrical particles is likewise determined by the Aspherical Shape characteristic. The X component stipulates the cylinder's radius, while the Z component specifies its length laterally to the local Z-axis. The Y component is usually overlooked. The orientation of aspherical particles is determined by the "Orientation" particle attribute, which consists of the four quaternionic components X, Y, Z, and W. If the Orientation attribute isn't defined or the four constituents will never form a unit quaternion, the aspherical particle stays aligned with the coordinate axes and is not rotated. To interpret the "Aspherical Shape" and "Orientation" features, simulation files are frequently utilized.

The quati, quatj, quatw, and other atom characteristics of LAMMPS must be mapped to the Orientation.X, Orientation.Y, Orientation.Z, and Orientation.W assets, respectively, of OVITO when importing the dump file. Additionally, Aspherical Shape.X, Aspherical Shape.Y, and Aspherical Shape.Z must be mapped to the shapex, shapey, and shapez columns in OVITO, respectively. Because the aforementioned dump command outputs these characteristics as c orient* and c shape*, OVITO is unable to automatically construct this mapping. Use the "Edit column mapping" button in the file import screen to manually map the columns after loading the dump file [15–17].

1.3.4 Animating parameters and the camera

With OVITO's advanced animation tools, the user can do more than just display a simulation trajectory—the user can also animate the virtual camera, visual elements, and modifiers.

1.3.4.1 Animation frames vs simulation frames

The duration of the current animation interval is displayed on the timeline under the viewport area of OVITO once a simulation sequence has been loaded. In OVITO, the animation frame numbers begin at zero. For instance, there are 101 video frames in an animation that ends at frame 100. The time units utilized by the animation timeline, simulation frames, and animation frames have meaning differences that should be understood.

After that, the user will have the ability to adjust how long the animation lasts. Extending the animation interval past the number of imported simulation frames will add extra static video frames at the conclusion of the generated animation.

Moreover, the External file panel's Configure trajectory playback dialogue allows how the frames of the simulation trajectory are transferred to animation frames. The Playback rate parameter can be used to provide a mapping that isn't one-to-one. A replay rate of 0.5, for example, spreads the simulation order over two times the animation frames. The simulation's snapshots will then be seen for two video frames in a row. In contrast, a playback rate of 2x compresses the simulation order to half of the frames. The produced animation will then skip every other frame.

1.3.4.2 Time-dependent formula animation

Quickly design basic animation effects with the Expression selection and Compute property modifiers. By applying mathematical formulae or Boolean expressions, the user may modify certain particle properties or choose subgroups of particles (e.g., position, color, transparency, etc.).

The user may utilize the unique variable Frame, which assesses the existing animation frame value, in these types of mathematical computations. Because of this, OVITO recalculates the findings on each animation frame and the formula becomes time-dependent anytime this variable is used in an expression.

1.3.4.3 Animation through key frames

OVITO also offers an animation framework that is key-based. It enables the user to set an animatable parameter's value at certain animation points, such as the start and finish. The parameter's value is then calculated in the intervals between these crucial frames using smooth interpolation, which is usually linear.

The input field's neighboring "A" button denotes animatable parameters. By clicking on the button, the animation key dialogue for that constraint appears, allowing the addition of more animation keys. Animation keys behave as time-value pairs. Two animation keys are present in the Distance parameter of the slicing plane in the screenshot on the right, one at frame 0 and the other at frame 50. As a result, by the second animation key frame, the parameter's value will have increased from 20.0 to 80.0. An animatable parameter will always have at minimum one animation key.

The constant value of the parameter is defined as long as that one key is present. When adding a new key, OVITO interpolates between the existing key values, causing the parameter to appear animated.

1.3.4.4 Auto-key mode

On the animation toolbar at the lowermost of the main window, there is a button with a key icon that triggers the "Auto-key" mode. Enabling this mode leads to backdrop of the time slider being highlighted in red, and each modification made to an animate-able parameter inside the program will produce a new animation key at the existing animation time. For example, after the Auto-key mode is established, the user can utilize the time slider to jump to animation frame 0 and then set the Slice modifier's Distance parameter to 20. Next, adjust the parameter value to 80 in the last frame of the animation. Every time one of the two parameters changes, a new animation key is created at the associated animation time.

The Auto-key mode should be disabled once designing the required animation keys is completed. If not, the danger of inadvertently making additional keys for parameters that the user wishes to leave unanimated is high. Creating new animation keys is quick and simple with the Auto-key mode, and it's frequently faster than using the animation key dialog.

1.3.4.5 Animating the camera

Before animating a camera object, the user must first construct it. This is accomplished using the viewport's context menu. The new camera object will be positioned in the three-dimensional scene at the current perspective. Additionally, when the camera is changed, the current viewport will be connected to the new camera object, enabling the viewport to update and display the scene as seen from the new camera location. The camera object in the scene will move if the user uses the zoom, pan, or orbit functions to change the viewport's perspective. The newly generated camera object appears in the other three viewports.

The "Move" and "Rotate" tools on the main toolbar may be used to move the camera object around in space. When either of these modes is chosen, the user may use the mouse to manipulate the camera object or enter the position coordinates and Euler orientation angles of the selected item using the numeric input boxes in the status bar area. The keyframing technique may be used to animate the camera object's position and orientation much like other characteristics.

Before the user may produce animation keys for the position or rotation of the camera, one must first enable the Auto-key mode. Animation keys are generated automatically when the user moves the camera at different animation timings; the camera travels an interpolated path between the keyframe points. The created animation keys are also shown in the track bar when the camera is selected.

Additionally, the user may animate the simulation box and have it spin in place of moving the camera's position. To do this similarly, select the simulation box and use the Rotate tool when the Auto-key mode is on [18, 19].

1.3.5 Pipeline cloning

Using OVITO's "Clone Pipeline" tool, the user may replicate the existing data pipeline and see different datasets side by side or in several views inside an image. The pipeline selector widget included in OVITO's main toolbar can be used to invoke the function. The user may now adjust the cloning process by using the Clone Pipeline interface. The cloning process is carried out by the parameters you choose in the dialogue when you hit OK to end it.

After the pipeline has been cloned, you will see two objects in the viewports: the new copy and the original data pipeline. These will both be a part of the current visualization scene. By clicking on an object in the interactive viewport, you may select the pipeline associated with that object.

1.3.5.1 Working of pipeline cloning

A pipeline in OVITO consists of a data source, which is often an external data file that feeds the pipeline with input, and a linear series of modifiers that have been added to the pipeline. In the OVITO user interface, the modifiers are often shown in a bottom-up manner, with the data source at the bottom and the final modifier in the sequence at the top. An input dataset, such as particles or other data items, is provided by the data source and is subsequently treated one by one by the modifiers. The information displayed in OVITO's 3D viewports is the data that is the final output of the pipeline.

The data source and/or modifiers from the original pipeline are often transferred to the new pipeline when cloning it. You can, however, indicate whether or not the original and cloned pipelines should share or duplicate each of these components independently. This gives you a great deal of freedom: For example, if you choose that the two pipelines should share the same data source, then the same input dataset will always be used by both processes. The other pipeline will also be impacted if you later decide to update the input file for one of the pipelines.

By employing different modifiers in the two pipelines that share the same data source, this synchronized mode, for instance, may be used to examine the same input data in several ways.

Seeing two distinct input datasets side by side while making sure that all modifier actions are done consistently to both datasets is another common scenario for pipeline cloning. In this case, you would split the modifiers across the two processes and replicate the data source during the cloning phase. After that, different input datasets will be fed into the pipelines with the guarantee that every dataset will undergo the same set of changes.

In the Clone Pipeline dialog, you can choose between four alternative cloning modes for any data source or modification of an existing pipeline:

Copy: This creates an independent clone of the original object. The other pipeline is unaffected by changes made to an object's parameters in either the original or cloned pipeline.

Join: The item will be shared by the new pipeline and the current pipeline; it will not be duplicated. As it turns out, selecting this option starts a branching pipeline from a single data source. Up to the branching point—the final modification that was cloned using mode Join—the data moving through the pipeline will be unique. After the branching point, the two pipeline branches will be autonomous and able to alter the data in different ways.

Share: This method can only be used to copy modifiers; it cannot copy the data source. It's a specific variation of the Join mode that provides additional choices. Both pipelines incorporate the common modifier, and changes made to their parameters will impact both pipelines in the same way. The shared modifier, however, can act on different data in the two pipelines since the two pipelines might have different sources and/or modifiers prior to the shared modifier, unlike a branching pipeline design, which is always restricted to a single unique data source.

Skip: All this option does in the cloned pipeline is delete a modifier.

1.3.5.2 After cloning a pipeline

Following a pipeline's cloning, the original and the cloned pipeline could share certain modifiers or the data source. Examine the pipeline example below, which is being copied by the schematic diagram and has four modifiers:

> The cloned pipeline uses the first modifier (cluster analysis) and data source of the original pipeline in accordance with the chosen mode Join. Consequently, the two pipelines' upstream segments will be shared. The first item to use a clone mode other than Join is the Slice modifier. Consequently, the cloned pipeline does not branch before this

modification, but the copied pipeline does. The selection modifier for Expression is in the Mode Copy state. Consequently, a duplicate of this alteration will be created and introduced into the replicated pipeline autonomously.

The freshly produced pipeline will show up in OVITO's pipeline editor after the cloning process is finished:

The horizontal "Pipeline branch" line shows that additional pipelines in the same situation share the space under the line with the current pipeline. These additional pipelines will be impacted by any modifications that are made, deleted, or altered below this line. Some of the list entries in the pipeline editor are displayed in italicized type. OVITO uses the italic typeface to indicate that a given item is shared by two or more pipelines. For example, the Assign color modifier was cloned via Mode Share. It is therefore a component of the original pipeline as well as the copies. Adjusting the color parameter of the Assign color modifier will result in a recalculation of both pipelines that utilize the modifier.

You could discover later on that choosing to share the Assign color modifier was a mistake. For instance, you can alter the chosen particles' color in the cloned pipeline while keeping the pipeline's original color. To do this, OVITO provides the Replace with independent copy function (highlighted in red). You may use it to make a unique copy of the standard Assign color modifier. Therefore, the other pipeline is unaffected by changes made to the modifier's parameters.

The visual elements of the pipeline editor are also shown in an italicized typeface. This is because these visual components are produced by the pipeline's data source, which is situated in the pipeline's upstream area that is shared by several pipelines. These visual components are shared objects, thus altering their settings or turning on or off their display will have an impact on how the datasets are shown in both pipelines. If required, you may duplicate the visual components and change the particle and bond visual appearances separately for the original and cloned pipelines using the "Replace with independent copy" tool in this situation [20].

1.3.6 Remote data access

OVITO has built-in SSH (Secure Shell) and HTTP(S) (HyperText Transfer Protocol Secure) clients for accessing files on distant workstations. This feature removes the requirement to first move files to your local desktop computer from remote locations, such as HPC (High Performance Computing) clusters [4, 21, 22]. For more details, visit Remote file access—OVITO User Manual 3.10.6 documentation.

1.4 REFERENCE SECTION

1.4.1 Pipelines

Data pipelines are the basis of the OVITO workflow. The sections that follow outline the many components that make up an OVITO data pipeline:

1.4.1.1 Data sources

As it supplies the input data that is sent to the modifiers, a data source is a crucial part of any pipeline. The pipeline editor's Data source section shows the data source for the current pipeline. The option that imports the pipeline's input data from one or more files on your local computer or a remote workstation is usually the one you'll select: external file [23]. To view the supported data sources, visit Data sources—OVITO User Manual 3.10.6 documentation.

1.4.1.2 Data objects

Many distinct data items, each representing a different component of the data (e.g., simulation cell form, atomic coordinates, bond list, etc.), may make up a dataset imported into OVITO from a simulation file. As the dataset passes through the data pipeline, modifiers can add dynamically calculated data objects to it. They operate with these data items. So-called visual components, responsible for creating three-dimensional graphical representations of numerical data stored in data objects, are assigned the task of visualizing data in OVITO. This indicates that, rather than the data items themselves, only the visual components that produce the graphical representation have changing visualization properties [24].

1.4.1.3 Visual elements

OVITO distinguishes between the underlying data and the resulting visual representations. This distinction between data items and their visual representation is well shown by the Position particle attribute, which preserves the XYZ coordinates of a collection of point-like particles. To show this data, OVITO automatically creates a Particles visual element, which uses the Position particle attribute as source data to render a corresponding set of spheres in the viewports. The glyph form, which can be spheres, disks, cubes, or other shapes, is one of the many features of the Particles visual element that determines how the particles appear visually.

The division of visual components and data items is advantageous to OVITO since it provides greater flexibility. It becomes possible to compare and contrast different datasets and present the same data in a multitude of ways (various graphic components based on the same source data) (one visual element rendering several data objects in the same way). Visual components are often automatically created by OVITO's data pipeline technology in addition to the imported or processed data. You may find them all under the "Visual elements" section of the pipeline editor. The pipeline editor's check boxes adjacent to each visual element can be used to disable it. As such, the generated pictures and related data visualization in interactive viewports will be deactivated [25].

1.4.1.4 Modifiers

The essential elements of an OVITO data pipeline are modifiers. Just like the components of a toolbox, each modifier carries out a very particular, well-defined operation or computation, and more complicated tasks typically need the combination of many modifiers [26]. To understand the modifiers in detail visit, Modifiers—OVITO User Manual 3.10.6 documentation

1.4.2 Viewports

1.4.2.1 Viewport layers

On top of the three-dimensional scene, viewport layers render two-dimensional text and images. They let you add more information to the photos and animations you produce. A viewport may be applied using a variety of layer types that OVITO offers. To manage the layers of the active viewport, navigate to the "Viewport layers" tab in the OVITO command panel. Only when a viewport's render preview mode is on are viewport layers displayed in the interactive viewport windows. OVITO instantly enters render preview mode when you add a new layer to the active viewport. For more details about available viewport layer types, visit Viewport layers—OVITO User Manual 3.10.6 documentation.

1.4.2.2 Adjust view dialog

This dialog box allows you to numerically enter the camera position and orientation of a viewport. It's accessible via the viewport menu.

View position: The virtual camera's Cartesian coordinates in space.
View direction: The direction vector of the virtual camera.
Up direction: The virtual camera's vertical direction vector in the projection plane. This vector is shown for reference only and cannot be modified in the dialogue.
Type of projection: When using a perspective projection, the camera's vertical field of vision is defined as a degree angle. The vertical field of view in a parallel projection is expressed in length units [27].

1.4.3 Rendering

Rendering is the process of producing excellent output photos of the current dataset for use in publications or other presentations. You may change a number of rendering parameters on the command panel's rendering page. Here, for example, you may select whether OVITO renders a single image, a series of photos, or a movie. Important parameters like the generated photos' backdrop color and output resolution are also modifiable. For more details about the rendering engine, visit Rendering—OVITO User Manual 3.10.6 documentation.

1.4.3.1 Render settings

General rendering options for images and movies, including resolution, output filename, backdrop color, and more, are available in this panel. It's on the Rendering tab of the command panel. Click the "Render active viewport" button in the toolbar above the viewports or the button at the bottom of the panel to start rendering a picture or video. The viewport that is now active, shown by a yellow border, is what OVITO renders by default. To set a viewport as the active one, right-click on it.

When rendering is finished, the produced pictures are seen in the frame buffer window; however, they are not immediately saved to the disk. As soon as you quit OVITO or shut the frame buffer window, the pictures will be gone. Use the "Save to file" button from the frame buffer window's toolbar when rendering is finished, or turn on the Save to file feature from the Render settings panel in advance. When rendering a movie, bear in mind that the Animation Settings dialog, which you can reach by selecting the "Animation settings" button, is where you define the animation's duration and frame rate (in frames per second).

Range of rendering: decides which frames of the animation should be rendered. Make sure to choose "Save to file" and provide a filename in the box below if you wish to render an animation, such as a sequence of frames. If not, the produced animation won't be permanently stored on disk.

Every Nth frame: The proportion of rendered animation frames is indicated by this parameter. All animation frames are rendered when this option is set to 1. Only every other frame will be rendered when the number is 2, and so on. As such, you may use this to control the number of generated video frames in the event of very long simulation sessions.

File number base: When an animation's frames are saved as a collection of picture files using OVITO, the frame number is automatically appended to the output filename. You can specify an alternate base file number, or offset, using this option. This comes in useful if you want to gather the video frames from many animation sequences into a single directory, which you can subsequently combine into a single continuous movie using external video encoding tools.

Output image size: This determines the width and height in pixels of the produced image or video. Interestingly, you can see precisely which rectangular portion of the viewport will display in produced photographs by using the preview mode for the specified viewport.

Save to file: This option directs OVITO to store the generated picture or video on disk in addition to displaying it in the frame buffer window. The "Choose..." option allows you to choose an output filename. When creating an animation sequence, all of the animation frames will be written to a single movie file if you select a video format like AVI or MPEG. OVITO will generate a series of image files, one for each frame, if you choose an image format (such as PNG or JPEG). In this case, the base filename you choose will automatically have the frame number attached to it.

Background: You can choose between a transparent backdrop or a background color that is consistent. In this case, OVITO will render visible those areas of the output picture that are not occupied by any objects. This is useful if you wish to utilize the image later on, say, on a presentation slide, in front of a different background. Keep in mind that for it to function, the output format has to allow transparency information. The ideal format for files is PNG.

Switch renderer: This option allows you to choose a different rendering engine. Beneath the Render settings panel is a second panel that shows the parameters and settings specific to each render engine [28].

1.4.3.2 Animation settings

From the animation toolbar, you can access the Animation Settings dialog. The dialogue allows you to change animation-related settings such as the playback frame rate and the current animation's length.

Frames per second: The frame rate of the animation is stated in frames per second. This option controls the frame rate of the animation played in interactive viewports and the video files produced by OVITO.

Playback speed: Controls the animation playback speed in OVITO's interactive viewports. This multiplier is multiplied by the above-mentioned frame rate (FPS) parameter. Note that if loading, processing, and displaying the next frame takes a lengthy time, the effective playback speed will be substantially slower.

Every Nth frame: This allows OVITO to skip some animation frames in interactive viewports during playback. This can be used to speed up the playback of a long trajectory when loading and computing each frame would take too long.

Loop playback: This controls whether or not the viewport's animation is played in a loop. If this option is disabled, playing will cease when the animation reaches its conclusion.

Custom animation interval: This option allows you to change the length of the animation interval displayed in OVITO's timeline. This interval is normally changed automatically based on the length of the loaded simulation sequence. Setting a custom animation duration is beneficial if the simulation dataset only contains a static frame (no trajectory), but you want to construct a camera animation that moves around the dataset [29, 30].

1.4.3.3 OpenGL renderer

This is the standard built-in rendering engine, which OVITO also uses to create interactive viewports.

Anti-aliasing level: The output image is frequently displayed at a higher resolution than the final image to reduce aliasing problems (super sampling). This parameter determines how much greater the resolution will be. Anti-aliasing is disabled by a factor of one. Higher values indicate a higher level of quality [31].

1.4.4 Data Inspector

The data inspector is a panel that appears beneath the viewport area and provides a detailed look at the data that has been dynamically created by the data pipeline or imported from an input file. The data inspector panel is initially collapsed, showing only the tab bar. You may make the tab bar larger by clicking anywhere inside it. There are several pages in the panel, and the pages are dynamically shown according to the kinds of data in the dataset that is now being used. For example, there is a website that has a tabular list of all particles and their property data.

1.4.4.1 Particles

The leftmost column displays the particle index number, which in OVITO begins at 0. For every particle attribute, the values are displayed in the following columns. By putting a Boolean expression in the input field located at the top of the table, you may dynamically filter the list of particles that is presented. Consider: In the filter field, enter "Coordination==11" to display only particles having a coordination number of 11. Logic AND and OR operators may be used to combine a wide range of criteria. The expression syntax used by the Expression selection modifier is also used by the Expression selection modifier. Clear the filter by pressing the X button to get back to the full list of particles.

When the crosshair button is in mouse input mode, you may choose certain particles inside the viewports. As you choose particles in the viewports, the filter expression is instantly changed to display the attributes of the highlighted particles. You may choose several particles by holding down the Ctrl key. Press the crosshair button once more or do a right-click in a viewport to end the input mode. To view a second table with particle distances, click the center tool button. Here, OVITO reports the pair-wise distances for the first four particles in the particle list.

To identify a group of two, three, or four particles for which the interparticle distances need to be calculated, filter the particle list. Note that this function ignores periodic boundary constraints when computing pair-wise distances. If you are interested in wrapped distances, you should establish linkages between the particles and measure the length of these connections rather than measuring the length of the ties between the particles.

1.4.4.2 Bonds

This data inspector page shows all bonds together with their property values in a data table. The page will appear if there are bonds in the current dataset. In the input section at the top of the table, you may add a Boolean expression to dynamically filter the bonds list that is presented. Examine how each bond's Length property was determined and applied using the Compute property modifier. To restrict the results to only show bonds longer than 1.5 Angstroms, type the phrase "Length 1.5" into the filter field.

Logic AND and OR operators may be used to combine a wide range of criteria. Click "X" to remove the filter and see the complete list of bonds. The Topology bond property contains the indices of the two particles that are connected by a bond. This information can be used to choose the bonds that are linked to a specific particle. For example, we could use the following filter expression to pick all bonds next to the initial particle:

```
Topology.A==0 || Topology.B==0
```

1.4.4.3 Attributes

This data inspector page lists every attribute value associated with the current dataset. Examples of global characteristics associated with a dataset include the number of chosen particles or the current simulation timestep. If attributes are dynamically generated by the data pipeline for each frame, they may have time-dependent values that change with animation time. Each attribute has its name and current value, both of which are listed in the data inspector's table. The data source of the data pipeline generates some properties, such as "SourceFile" and "SourceFrame," while loading the input data from an external file. To convey calculation results, certain modifiers in the data pipeline may dynamically construct other properties.

The "file export" feature in OVITO can be used to export all or a subset of the table's characteristics to a text file. By pressing the save button on the right side of the data inspector, you may easily reach the "file export."

1.4.4.4 Data tables

This tab of the data inspector panel shows the data tables and charts generated by the current pipeline. A collection of one or more data columns that may be shown visually as a scatter plot, graph, or histogram is called a data table. Several data tables may be produced by the current data pipeline, which you may navigate between with the list box on the left. The tab's main portion displays the table that was selected. The tool buttons on the right allow you to alternate between the tabular view of the underlying numerical table data and the graphical plot mode.

The save button allows you to save the diagram as a PDF or PNG graphic file, or the numerical values as a text file if the tabular view is now active.

Mouse navigation is supported in the plot area:

- Left-click: show plot coordinates at the cursor position
- Left button + drag: zoom in to rectangular region
- Left button + Shift + drag: panning
- Mouse wheel: zoom in/out
- Right-click: reset view

1.4.4.5 Voxel grids

The voxel grids generated or loaded by the current data pipeline are listed in this tab of the data inspector panel.

1.4.4.6 Dislocations

A table of all dislocation lines in the current dataset can be found on this page of the data inspector. The Dislocation Analysis (DXA) modifier can be used to extract dislocation lines from an atomistic crystal. The mouse input mode is activated by pressing the crosshair button above the table, which allows you to choose particular dislocation lines in the viewports. The matching rows in the table are highlighted when you choose lines in the viewports. To select numerous dislocation lines, hold down the Ctrl key. This also works in the opposite direction: You can choose table rows in the highlighting mode, and OVITO will highlight the associated dislocations in the viewports. Press the crosshair button once more or do a right-click in a viewport to end the highlight mode [32–34].

1.4.5 Application settings

Select "Edit ‣ Application settings" from the main menu on Windows/Linux to launch the application settings window. Select "Ovito ‣ Preferences" on macOS. There are several tabs in the dialog:

- General settings
- Viewport settings
- Modifier templates
- Particle-related settings

1.4.5.1 General settings

The application settings dialog's tab allows you to modify global settings that impact the program's visuals and user interface.

1.4.5.1.1 User interface

Load file: Use alternative file selection dialog
This option modifies the file selector that displays when you load a simulation file into OVITO. By default, the file selection dialog box of the operating system is utilized. Rather than using the standard file selection, this option opens an integrated window that could or might not have any advantages.

Modifiers list: Sort by category
If this option is selected, the list of potential modifiers is separated into function groups. If not, modifiers are included in a single list alphabetically.

1.4.5.1.2 3D graphics

Graphics hardware interface
Select the application programming interface that will be used by OVITO to render interactive viewport contents. Currently, OVITO is compatible with Vulkan and OpenGL interfaces. Since OpenGL is a more widely used interface, it ought to function flawlessly on most systems. Vulkan is a more recent programming interface, although it may still have compatibility problems with some graphics drivers.

One benefit of the Vulkan interface is that if your system has several GPUs (Graphics Processing Unit) or integrated graphics processors, you may select which one OVITO will use directly. When utilizing the OpenGL interface, this decision needs to be made at the operating system or graphics driver level. To gain further knowledge about your system's graphics hardware, navigate to OVITO's "Help" menu and choose "System Information." When reporting problems with graphics compatibility, provide the OVITO developers this information.

1.4.5.1.3 Program updates

Auto-refresh news page from web server
If software updates are enabled, the application will connect to the OVITO server at ovito.org each time it launches in order to check for updates. The command panel of the application indicates the availability of a new update.

Send unique installation ID to web server
In order to help the program maker gather anonymous usage information, OVITO transmits an installation ID to the OVITO server during the update check if this option is chosen. The installation ID is a unique, anonymous, random number that is produced during the local software installation. It is devoid of any user or personal information. Furthermore, OVITO GmbH, the software developer, does not gather or retain any information about you, save for the anonymous installation ID that may be used to identify you. OVITO will keep operating as usual even if this option is disabled.

1.4.5.2 Viewport settings

The interactive viewports of the OVITO are covered on this page of the application settings dialog.

1.4.5.2.1 Camera

Coordinate system orientation
OVITO can limit the rotation of the viewport camera so that the chosen Cartesian coordinate axis always points upward. The Z-axis is the default.

Restrict camera rotation to keep the major axis pointing upward
This setting prevents the camera from turning upside down by constraining its orientation.

1.4.5.2.2 Color scheme

This option lets you choose between a dark and a white viewport background.

1.4.5.3 Modifier templates

Modifier templates allow you to create modifiers with pre-configured settings for easy access. This is handy in instances where you need to apply the same modifier on multiple datasets with the same settings. The modifier templates you create are saved across program sessions and can even be moved from one machine to another. Furthermore, a modifier template can contain many modifiers, allowing you to quickly insert frequently used sequences or combinations of modifiers into the data pipeline.

The application settings dialogue is where you manage the modifier templates. The pipeline editor has a button for rapidly accessing that dialogue. All modifier templates you create are available across the entire application, and OVITO remembers them across program sessions. Existing modifiers in the active data pipeline are used to define new templates. That is, you "store" one or more modifiers, including their current settings, so they can be used again later. Modifier templates appear behind the ordinary modifiers in the drop-down list of possible modifiers once they've been defined. They may be simply applied to fresh datasets from here. If you need to apply the same modifier with the same settings over and over, this feature comes in handy.

User-defined script modifiers are another common application for modifier templates. You can use the Python script modifier to create a script function that does user-defined data operations. After you've created a new script function, you can save it as a modifier template and use it in future program sessions, exactly like the normal OVITO modifiers. Because the script code will be part of the stored template, you won't have to save it to an external source file and manually load it into OVITO.

In a proprietary format, OVITO saves the definitions of all modifier templates in the application's settings file. They can be exported to a special modification template file if necessary, so they can be transferred to another computer and imported into another OVITO installation.

1.4.5.4 Particle-related settings

This page of the application settings dialog allows you to modify the program's defaults for particle sizes and colors. OVITO uses the particle types' names to determine standard colors and radii when importing a simulation file. Particularly for well-known chemical elements like "He," "Fe," and "Si," this is accurate. The table shows the predefined relationships between the designated particle types and their display radii, default colors, and van der Waals radii. You can alter the settings if needed. OVITO will retain these default settings for newly imported datasets and apply them to them over program sessions.

Normally, after importing a simulation file, you may change the look of particle types from the Particle types panel, where you can also set new default settings for certain particle types. This is also where you may customize color and radius parameters for particle types other than the standard chemical components. All user-defined presets will be deleted and all colors and radii will be set back to OVITO's hard-coded factory default settings upon selecting the "Restore built-in defaults" option.

1.4.5.5 *Where does OVITO store its settings?*

OVITO stores user preferences in a place that varies depending on the platform in order to maintain them across program sessions. On Windows, the data is stored in the system registry. On Linux and macOS, it is stored as a text-based configuration file in the user's home directory. The precise location of the machine's storage is shown at the bottom of the program settings dialog box. To return OVITO to factory default settings, just use the Windows Registry Editor tool to delete the registry branch on Windows, or erase the configuration file on Linux/macOS [35].

In addition, the authors have extensive backgrounds in the experimental characterization of composite materials and molecular dynamics [36–93].

ACKNOWLEDGMENT

Monetary and academic support from the University of Petroleum and Energy Studies, Dehradun, India (SEED Grant program), is highly appreciable.

CONFLICTS OF INTEREST

"There are no conflicts of interest to declare by the authors."

REFERENCES

[1] Stukowski, A. 2009. Visualization and analysis of atomistic simulation data with OVITO—The open visualization tool. *Modelling and Simulation in Materials Science and Engineering*, *18*(1), 015012.

[2] Weng, S., Chen, X., Yue, X., Fu, T. and Peng, X. 2019. Inapparent strengthening effect of twin interface in Cu/Pd multilayered films with a large lattice mismatch. *Nanomaterials*, 9(12), 1778.

[3] Ovito.org. 2021. *About OVITO — OVITO user manual 3.5.4 documentation* [online] Available at: https://www.ovito.org/docs/current/introduction.html.

[4] www.ve4s.ru. n.d. [online] Available at: http://www.ve4s.ru/lammps-manual-dumpling.html.

[5] www.ovito.org. n.d. *Data import — OVITO user manual 3.5.4 documentation* [online] Available at: https://www.ovito.org/docs/current/usage/import.html.

[6] Ovito.org. 2021. *Viewport windows — OVITO user manual 3.5.4 documentation* [online] Available at: https://www.ovito.org/docs/current/usage/viewports.html.

[7] Ovito.org. 2021. *Data model — OVITO user manual 3.5.4 documentation* [online] Available at: https://ovito.org/manual_testing/usage/data_model.html

[8] LaptrinhX 2019. *Performance of pipeline architecture: The impact of the number of workers* [online] LaptrinhX. Available at: https://laptrinhx.com/performance-of-pipeline-architecture-the-impact-of-the-number-of-workers-2731519165/

[9] Ovito.org. 2021. *Pipeline concept — OVITO user manual 3.5.4 documentation* [online] Available at: https://ovito.org/manual_testing/usage/pipeline.html
[10] Ovito.org. 2021. *Rendering — OVITO user manual 3.5.4 documentation* [online] Available at: https://ovito.org/manual_testing/usage/rendering.html
[11] Ovito.org. 2021. *Miscellaneous — OVITO user manual 3.5.4 documentation* [online] Available at: https://ovito.org/manual_testing/usage/miscellaneous.html
[12] Ovito.org. 2021. *Exporting data — OVITO user manual 3.5.4 documentation* [online] Available at: https://ovito.org/manual_testing/usage/export.html
[13] Ovito.org. 2021. *Semi-transparent particles — OVITO user manual 3.5.4 documentation* [online] Available at: https://ovito.org/manual_testing/advanced_topics/transparent_particles.html
[14] Ovito.org. 2021. *Adding a scale bar to a viewport — OVITO user manual 3.5.4 documentation* [online] Available at: https://ovito.org/manual_testing/advanced_topics/scale_bar.html
[15] Nguyen, T.D. and Plimpton, S.J. 2019. Aspherical particle models for molecular dynamics simulation. *Computer Physics Communications*, *243*, 12–24.
[16] Lammps.org. 2021. *Virtual LAMMPS workshop and symposium* [online] Available at: https://www.lammps.org/workshops/Aug21/tutorial/
[17] Ovito.org. 2021. *Non-spherical particle shapes — OVITO user manual 3.5.4 documentation* [online] Available at: https://www.ovito.org/docs/current/advanced_topics/aspherical_particles.html
[18] www.ovito.org. n.d. *Animating parameters and the camera — OVITO user manual 3.5.4 documentation* [online] Available at: https://www.ovito.org/docs/current/advanced_topics/animation.
[19] Android Developers. n.d. *Camera API* [online] Available at: https://developer.android.com/guide/topics/media/camera.
[20] www.ovito.org. n.d. *Pipeline cloning — OVITO user manual 3.5.4 documentation* [online] Available at: https://www.ovito.org/docs/current/advanced_topics/clone_pipeline.html
[21] www.ovito.org. n.d. *Remote data access — OVITO user manual 3.5.4 documentation* [online] Available at: https://www.ovito.org/docs/current/advanced_topics/remote_file_access.html#usage-import-remote
[22] www.pixelstech.net. n.d. *Using public key authentication in SSH* [online] Available at: https://www.pixelstech.net/article/1473500460-Using-public-key-authentication-in-SSH
[23] www.ovito.org. n.d. *Data sources — OVITO user manual 3.5.4 documentation* [online] Available at: https://www.ovito.org/docs/current/reference/pipelines/data_sources/index.html
[24] www.ovito.org. n.d. *Data objects — OVITO user manual 3.5.4 documentation* [online] Available at: https://www.ovito.org/docs/current/reference/pipelines/data_objects/index.html
[25] www.ovito.org. n.d. *Visual elements — OVITO user manual 3.5.4 documentation* [online] Available at: https://www.ovito.org/docs/current/reference/pipelines/visual_elements/index.html].
[26] www.ovito.org. n.d. *Modifiers — OVITO user manual 3.5.4 documentation* [online] Available at: https://www.ovito.org/docs/current/reference/pipelines/modifiers/index.html
[27] www.ovito.org. n.d. *Viewport layers — OVITO user manual 3.5.4 documentation* [online]. Available at: https://www.ovito.org/docs/current/reference/viewports/layers/index.html

[28] www.ovito.org. n.d. *Render settings — OVITO user manual 3.5.4 documentation* [online]. Available at: https://www.ovito.org/docs/current/reference/rendering/render_settings.html

[29] autodesk.com. n.d. *3ds max reference* [online] Available at: https://download.autodesk.com/us/3dsmax/2012help/index.html?url=files/GUID-9473CAF3-AF73-4127-A98C-58ACEF01ACA-3023.htm

[30] www.ovito.org. n.d. *Animation settings—OVITO user manual 3.5.4 documentation* [online]. Available at: https://www.ovito.org/docs/current/reference/rendering/animation_settings_dialog.html

[31] Chen, B. and Cheng, H.H. 2005. Interpretive OpenGL for computer graphics. *Computers & Graphics*, 29(3), 331–339.

[32] www.ovito.org. n.d. *Data inspector — OVITO user manual 3.5.4 documentation* [online]. Available at: https://www.ovito.org/docs/current/reference/data_inspector/index.html

[33] www.ovito.org. n.d. *ovito.io — OVITO scripting reference 3.5.4 documentation* [online]. Available at: https://www.ovito.org/docs/current/python/modules/ovito_io.html

[34] Stukowski, A. 2014. A triangulation-based method to identify dislocations in atomistic models. *Journal of the Mechanics and Physics of Solids*, *70*, 314–319.

[35] docs.ovito.org. n.d. *General settings — OVITO user manual 3.5.4 documentation* [online] Available at: https://docs.ovito.org/reference/app_settings/general_settings.html

[36] Verma, A., Parashar, A. and Packirisamy, M. 2018. Atomistic modeling of graphene/hexagonal boron nitride polymer nanocomposites: A review. *Wiley Interdisciplinary Reviews: Computational Molecular Science*, *8*(3), e1346.

[37] Verma, A., Singh, V.K., Verma, S.K. and Sharma, A. 2016. Human hair: a biodegradable composite fiber—A review. *International Journal of Waste Resources*, *6*(206), 2.

[38] Verma, A. and Singh, V.K. 2019. Mechanical, microstructural and thermal characterization of epoxy-based human hair–reinforced composites. *Journal of Testing and Evaluation*, *47*(2), 1193–1215.

[39] Verma, A. and Parashar, A. 2018. Structural and chemical insights into thermal transport for strained functionalised graphene: a molecular dynamics study. *Materials Research Express*, *5*(11), 115605.

[40] Verma, A., Negi, P. and Singh, V.K. 2019. Experimental analysis on carbon residuum transformed epoxy resin: Chicken feather fiber hybrid composite. *Polymer Composites*, *40*(7), 2690–2699.

[41] Verma, A., Gaur, A. and Singh, V.K. 2017. Mechanical properties and microstructure of starch and sisal fiber biocomposite modified with epoxy resin. *Materials Performance and Characterization*, *6*(1), 500–520.

[42] Verma, A., Parashar, A. and Packirisamy, M. 2019. Effect of grain boundaries on the interfacial behaviour of graphene-polyethylene nanocomposite. *Applied Surface Science*, *470*, 1085–1092.

[43] Verma, A., Budiyal, L., Sanjay, M.R. and Siengchin, S. 2019. Processing and characterization analysis of pyrolyzed oil rubber (from waste tires)-epoxy polymer blend composite for lightweight structures and coatings applications. *Polymer Engineering & Science*, *59*(10), 2041–2051.

[44] Verma, A., Negi, P. and Singh, V.K. 2018. Physical and thermal characterization of chicken feather fiber and crumb rubber reformed epoxy resin hybrid composite. *Advances in Civil Engineering Materials*, *7*(1), 538–557.

[45] Verma, A., Negi, P. and Singh, V.K. 2018. Experimental investigation of chicken feather fiber and crumb rubber reformed epoxy resin hybrid composite: mechanical and microstructural characterization. *Journal of the Mechanical Behavior of Materials*, 27(3–4).

[46] Chaurasia, A., Verma, A., Parashar, A. and Mulik, R.S. 2019. Experimental and computational studies to analyze the effect of h-BN nanosheets on mechanical behavior of h-BN/polyethylene nanocomposites. *The Journal of Physical Chemistry C*, *123*(32), 20059–20070.

[47] Jain, N., Verma, A. and Singh, V.K. 2019. Dynamic mechanical analysis and creep-recovery behaviour of polyvinyl alcohol based cross-linked biocomposite reinforced with basalt fiber. *Materials Research Express*, 6(10), 105373.

[48] Verma, A., Joshi, K., Gaur, A. and Singh, V.K. 2018. Starch-jute fiber hybrid biocomposite modified with an epoxy resin coating: fabrication and experimental characterization. *Journal of the Mechanical Behavior of Materials*, 27(5–6).

[49] Verma, A., Kumar, R. and Parashar, A. 2019. Enhanced thermal transport across a bi-crystalline graphene–polymer interface: An atomistic approach. *Physical Chemistry Chemical Physics*, *21*(11), 6229–6237.

[50] Verma, A. and Singh, V.K. 2016. Experimental investigations on thermal properties of coconut shell particles in DAP solution for use in green composite applications. *Journal of Materials Science and Engineering*, 5(3), 1000242.

[51] Verma, A., Singh, V.K. and Arif, M. 2016. Study of flame retardant and mechanical properties of coconut shell particles filled composite. *Research and Reviews: Journal of Material Sciences*, 4(3), 1–5.

[52] Verma, A., Parashar, A., Jain, N., Singh, V.K., Rangappa, S.M. and Siengchin, S. 2020. Surface modification techniques for the preparation of different novel biofibers for composites. In *Biofibers and biopolymers for biocomposites* (pp. 1–34). Springer.

[53] Rastogi, S., Verma, A. and Singh, V.K. 2020. Experimental response of nonwoven waste cellulose fabric–reinforced epoxy composites for high toughness and coating applications. *Materials Performance and Characterization*, 9(1), 151–172.

[54] Bharath, K.N., Madhu, P., Gowda, T.G., Verma, A., Sanjay, M.R. and Siengchin, S. 2020. A novel approach for development of printed circuit board from biofiber based composites. *Polymer Composites*, *41*(11), 4550–4558.

[55] Verma, A., Jain, N., Parashar, A., Gaur, A., Sanjay, M.R. and Siengchin, S. 2021. Lifecycle assessment of thermoplastic and thermosetting bamboo composites. In *Bamboo fiber composites* (pp. 235–246). Springer.

[56] Singh, K., Jain, N., Verma, A., Singh, V.K. and Chauhan, S. 2020. Functionalized graphite–reinforced cross-linked poly (vinyl alcohol) nanocomposites for vibration isolator application: morphology, mechanical, and thermal assessment. *Materials Performance and Characterization*, 9(1), 215–230.

[57] Verma, A. and Parashar, A. 2018. Molecular dynamics based simulations to study the fracture strength of monolayer graphene oxide. *Nanotechnology*, *29*(11), 115706.

[58] Verma, A. and Parashar, A. 2017. The effect of STW defects on the mechanical properties and fracture toughness of pristine and hydrogenated graphene. *Physical Chemistry Chemical Physics*, *19*(24), 16023–16037.

[59] Verma, A., Parashar, A. and Packirisamy, M. 2018. Tailoring the failure morphology of 2D bicrystalline graphene oxide. *Journal of Applied Physics*, *124*(1), 015102.

[60] Singla, V., Verma, A. and Parashar, A. 2018. A molecular dynamics based study to estimate the point defects formation energies in graphene containing STW defects. *Materials Research Express*, 6(1), 015606.
[61] Verma, A. and Parashar, A. 2018. Molecular dynamics based simulations to study failure morphology of hydroxyl and epoxide functionalised graphene. *Computational Materials Science*, *143*, 15–26.
[62] Verma, A., Zhang, W. and van Duin, A.C. 2021. ReaxFF reactive molecular dynamics simulations to study the interfacial dynamics between defective h-BN nanosheet and water nanodroplets. *Physical Chemistry Chemical Physics*, *23*, 10822–10834.
[63] Verma, A., Parashar, A. and Packirisamy, M. 2019. Role of chemical adatoms in fracture mechanics of graphene nanolayer. *Materials Today: Proceedings*, *11*, pp.920–924.
[64] Verma, A. and Parashar, A. 2020. Characterization of 2D nanomaterials for energy storage. In *Recent advances in theoretical, applied, computational and experimental mechanics* (pp. 221–226). Springer.
[65] Verma, A., Jain, N., Parashar, A., Singh, V.K., Sanjay, M.R. and Siengchin, S. 2020. Design and modeling of lightweight polymer composite structures. In *Lightweight polymer composite structures* (pp. 193–224). CRC Press.
[66] Verma, A., Jain, N., Parashar, A., Singh, V.K., Sanjay, M.R. and Siengchin, S. 2020. Design and modeling of lightweight polymer composite structures. In *Lightweight polymer composite structures* (pp. 193–224). CRC Press.
[67] Verma, A., Jain, N., Rastogi, S., Dogra, V., Sanjay, S.M., Siengchin, S. and Mansour, R. 2020. Mechanism, anti-corrosion protection and components of anti-corrosion polymer coatings. In *Polymer coatings: Technologies and applications* (pp. 53–66). CRC Press.
[68] Verma, A., Jain, N., Rastogi, S., Dogra, V., Sanjay, S.M., Siengchin, S. and Mansour, R. 2020. Mechanism, anti-corrosion protection and components of anti-corrosion polymer coatings. In *Polymer coatings: Technologies and applications* (pp. 53–66). CRC Press.
[69] Verma, A., Jain, N., Kalpana, Rangappa, S.M., Siengchin, S. and Jawaid, M. 2021. Natural fibers based bio-phenolic composites. *Phenolic polymers based composite materials*, pp.153–168.
[70] Bharath, K.N., Madhu, P., Gowda, T.Y., Verma, A., Sanjay, M.R. and Siengchin, S., 2021. Mechanical and Chemical Properties Evaluation of Sheep Wool Fiber–Reinforced Vinylester and Polyester Composites. *Materials Performance and Characterization*, *10*(1), pp.99–109.
[71] Marichelvam, M.K., Manimaran, P., Verma, A., Sanjay, M.R., Siengchin, S., Kandakodeeswaran, K. and Geetha, M., 2021. A novel palm sheath and sugarcane bagasse fiber based hybrid composites for automotive applications: An experimental approach. *Polymer Composites*, *42*(1), pp.512–521.
[72] Chaudhary, A., Sharma, S. and Verma, A., 2021. Optimization of WEDM process parameters for machining of heat treated ASSAB'88 tool steel using Response surface methodology (RSM). *Materials Today*: *Proceedings*.
[73] Chaudhary, A., Sharma, S. and Verma, A., 2021. WEDM machining of heat treated ASSAB '88 tool steel: A comprehensive experimental analysis. *Materials Today*: *Proceedings*.
[74] Verma, A. and Singh, V.K., Experimental characterization of modified epoxy resin assorted with almond shell particles. *ESSENCE-International Journal for Environmental Rehabilitation and Conservation*, p.36.

[75] Verma, A. and Samant, S.S., 2016. Inspection of hydrodynamic lubrication in infinitely long journal bearing with oscillating journal velocity. *Journal of Applied Mechanical Engineering*, *5*(3), pp.1–7.
[76] Verma, A., Parashar, A., Jain, N., Singh, V.K., Rangappa, S.M. and Siengchin, S., 2020. Surface modification techniques for the preparation of different novel biofibers for composites. *Biofibers and biopolymers for biocomposites*, pp.1–34.
[77] Bisht, N., Verma, A., Chauhan, S. and Singh, V.K., 2021. Effect of functionalized silicon carbide nano-particles as additive in cross-linked PVA based composites for vibration damping application. *Journal of Vinyl and Additive Technology*, 27(4), pp.920–932.
[78] Kataria, A., Verma, A., Sanjay, M.R. and Siengchin, S., 2021. Molecular modeling of 2D graphene grain boundaries: Mechanical and fracture aspects. *Materials Today: Proceedings*.
[79] Arpitha, G.R., Verma, A., Sanjay, M.R. and Siengchin, S., 2021. Preparation and Experimental Investigation on Mechanical and Tribological Performance of Hemp-Glass Fiber Reinforced Laminated Composites for Lightweight Applications. *Advances in Civil Engineering Materials*, 10 (1), 427–439.
[80] Deji, R., Verma, A., Kaur, N., Choudhary, B.C. and Sharma, R.K., 2021. Density functional theory study of carbon monoxide adsorption on transition metal doped armchair graphene nanoribbon. *Materials Today: Proceedings*.
[81] Deji, R., Verma, A., Choudhary, B.C. and Sharma, R.K., 2021. New insights into NO adsorption on alkali metal and transition metal doped graphene nanoribbon surface: A DFT approach. *Journal of Molecular Graphics and Modelling*. https://doi.org/10.1016/j.jmgm.2021.108109
[82] Kataria, A., Verma, A., Sethi, S.K. and Ogata, S., 2022. Introduction to interatomic potentials/forcefields. In Forcefields for Atomistic-Scale Simulations: Materials and Applications (pp. 21–49). Singapore: Springer Nature Singapore.
[83] Chaturvedi, S., Kataria, A., Chaudhary, V., Verma, A., Jain, N., Sanjay, M.R. and Siengchin, S., 2023. Bionanocomposites reinforced with cellulose fibers and agro-industrial wastes. In Cellulose fibre reinforced composites (pp. 317–342). Woodhead Publishing.
[84] Goswami, T., Bheemaraju, A., Kataria, A., Nag, A., Sravani, K., Mishra, S. and Mishra, A.K., 2021. Highly fluorescent water-soluble PTCA incorporated silver nano-cluster for sensing of dopamine. Materials Chemistry and Physics, 259, p.124086.
[85] Kataria, A., Chaturvedi, S., Chaudhary, V., Verma, A., Jain, N., Sanjay, M.R. and Siengchin, S., 2023. Cellulose fiber-reinforced composites—history of evolution, chemistry, and structure. In *Cellulose fibre reinforced composites* (pp. 1–22). Woodhead Publishing.
[86] Kataria, A., Verma, A., Sanjay, M.R., Siengchin, S. and Jawaid, M., 2022. Physical, morphological, structural, thermal, and tensile properties of coir fibers. In *Coir fiber and its composites* (pp. 79–107). Woodhead Publishing.
[87] Goswami, T., Kumar, S., Bheemaraju, A., Reddy, K.M., Sharma, A.K., Kataria, A. and Shrivastav, A., 2023. TiO2 Nanoparticles and Nb2O5 Nanorods Immobilized rGO for Efficient Visible-Light Photocatalysis and Catalytic Reduction. *Catalysis Letters*, *153*(2), pp.605–621.
[88] Goswami, Tapas, Navpreet Kamboj, Amarnath Bheemaraju, Aditya Kataria, and Ramendra Sundar Dey 2021. Supercapacitive behaviour of a novel

nanocomposite of 3, 4, 9, 10-perylenetetracarboxylic acid incorporated captopril-Ag nanocluster decorated on graphene nanosheets. *Materials Advances* 2(4), 1358–1368.
[89] Kataria, A., Liang, Z. and Dumitrica, T. 2023. Computational investigation of interacting nanohole defects for carbon nanotube cross-linking and functionalization with bismaleimide: Implications for ultrastrong, lightweight nanocomposites. *ACS Applied Nano Materials*, 6(8), 6474–6479.
[90] Kataria, A. and Verma, A. 2024. Continuum mechanics-based simulations to model creep recovery behavior of polymer composites. In *Dynamic mechanical and creep-recovery behavior of polymer-based composites* (pp. 311–335). Elsevier.
[91] Kataria, A., Wang, Y., Mujumdar, K. and Dumitrică, T. 2023. Influence of polymeric and cross-linked nano-hole interfaces on the tensile response of carbon nanotube bundles: A mesoscopic distinct element method approach. *Extreme Mechanics Letters*, 65, 102108.
[92] Kataria, A., Verma, A., Sharma, S., Mavinkere Rangappa, S. and Siengchin, S. 2023, February. molecular energies of lightweight Al, Cu and alloys: Evaluation and insights. In *International Symposium on Lightweight and Sustainable Polymeric Materials* (pp. 305–319). Singapore: Springer Nature Singapore.
[93] Kataria, A., Zafar, S., Verma, A. and Ogata, S., 2023. Molecular dynamics simulations in coatings. In *Coating Materials: Computational Aspects, Applications and Challenges* (pp. 125–156). Singapore: Springer Nature Singapore.

Chapter 2

Information processing and analysis through data visualization and data mining

Ajit Kumar Pasayat, Prashant Upadhyay and Sourajit Behera

2.1 INTRODUCTION

In the contemporary digital era, the sheer volume of data generated by various sources necessitates sophisticated methods for processing and analyzing this information. Data mining and data visualization emerge as pivotal techniques in this context, transforming raw data into meaningful insights (Myatt and Johnson, 2009). Data mining involves the application of advanced algorithms to uncover hidden patterns, trends, and relationships within vast datasets, which conventional methods might lack (Shu and Ye, 2023). This process enables businesses and researchers to make informed decisions, predict future trends, and enhance operational efficiencies. Concurrently, data visualization provides a visual context to data, making complex information more accessible and comprehensible (Qin et al., 2020). Through graphical representations such as charts, graphs, and plots, data visualization simplifies the interpretation of data analysis results, facilitating quicker and more effective decision-making. Together, these techniques are indispensable tools in fields ranging from market analysis and customer behavior prediction to educational outcomes and real-time data interaction, underscoring their significance in harnessing the potential of big data (Allil, 2024). This chapter delves into the methodologies, applications, and advancements in data mining and visualization, illustrating their critical role in modern information processing and analysis.

2.2 UNDERSTANDING DATA VISUALIZATION

2.2.1 Definition and importance

Data visualization is the graphical representation of information and data. By utilizing visual elements like charts, graphs, and maps, data visualization tools provide an accessible way to see and understand trends, outliers, and patterns in data (Luo et al., 2019). The primary goal of data visualization is to communicate information clearly and efficiently to users through

 DOI: 10.1201/9781003621102-2

graphical means. This process not only makes complex data more understandable but also aids in the quick identification of patterns and insights that might be missed in text-based data.

The importance of data visualization lies in its ability to make data more comprehensible and actionable. Visualization techniques transform raw data into a visual context that helps users grasp difficult concepts or identify new patterns. For example, in education, robust data visualization methods, such as raincloud plots, can significantly enhance the understanding and interpretation of learning outcomes, facilitating better decision-making and strategy development. In business, data visualization tools can predict customer behavior, understand product usage, and create forecasting models for specific markets, thereby helping organizations make data-driven decisions.

Data visualization is essential because it enables users to interact with data, explore different scenarios, and make informed decisions based on visual evidence. It transforms complex datasets into intuitive visual formats that can be easily understood by a broader audience, fostering a data-driven culture within organizations and across various fields.

2.2.2 Principles of effective data visualization (Midway, 2020)

Effective data visualization hinges on several key principles:

- **Clarity:** Visuals should be easy to understand, ensuring that the audience can quickly grasp the underlying message without confusion. Clear visuals eliminate ambiguity by using straightforward and unambiguous graphical representations. This can involve the use of simple charts, appropriate labeling, and avoiding clutter that can obscure the data.
- **Accuracy:** Representations must reflect the true nature of the data. It is crucial that visualizations do not distort the data, as misleading visuals can result in incorrect interpretations and decisions. This includes choosing the correct type of visualization (e.g., bar charts, line graphs) and ensuring that scales, axes, and labels are used appropriately to convey the true meaning of the data.
- **Efficiency:** Data should be conveyed in a manner that can be quickly and accurately interpreted. Efficient visualizations allow viewers to extract insights rapidly, focusing on the most relevant data points and eliminating unnecessary elements that could overwhelm the audience. This can involve the use of visual hierarchies, concise legends, and avoiding the overuse of colors or patterns that may distract from the key information.
- **Aesthetics:** Good design enhances comprehension and retention. Aesthetically pleasing visualizations are more engaging and can help in better retention of information. This involves using color, layout, and design principles effectively to make the visualization not only

functional but also appealing. Well-designed visuals draw the viewer's attention to the most important aspects of the data and help in maintaining interest and engagement.

By adhering to these principles, data visualization can effectively communicate information, support decision-making, and drive insights in a clear, accurate, efficient, and visually appealing manner. These principles ensure that data visualizations serve their purpose of making complex information accessible and understandable, facilitating better decision-making and enhanced data interpretation.

2.2.3 Types of data visualizations

2.2.3.1 Various types of visualizations and their purposes

- Bar charts (Talbot et al., 2014):
  - **Purpose:** Bar charts are ideal for comparing quantities across different categories as represented in Figure 2.1. They use rectangular bars with lengths proportional to the values they represent, making it easy to compare data points directly.
  - **Details:** Bar charts are particularly effective when the data involves discrete categories. They can be displayed either vertically or horizontally. This type of chart is commonly used in business and marketing to compare sales figures, survey responses, or other categorical data.
- Line Charts(Lal et al., 2023):
  - **Purpose:** Line charts are useful for showing trends over time as represented in Figure 2.2. They display data points connected by straight lines, which makes it easy to observe how values change at regular intervals.
  - **Details:** Line charts are ideal for time-series data, such as stock prices, temperature records, or any other data that changes continuously over time. They help in identifying trends, patterns, and fluctuations, making them crucial for financial and scientific data analysis.
- Pie Charts (Siirtola, 2019):
  - **Purpose:** Pie charts are best for displaying proportional data as represented in Figure 2.3. They show the parts of a whole, with each slice representing a category's contribution to the total.
  - **Details:** Pie charts are effective for showing the composition of a dataset. Each slice's angle and area are proportional to the quantity it represents, making it useful for displaying data like budget allocations, market share, or demographic compositions.
- Scatter Plots (Keim et al., 2010):
  - **Purpose:** Scatter plots are good for identifying relationships between variables as represented in Figure 2.4. They plot data points on a Cartesian plane, which helps in visualizing correlations.

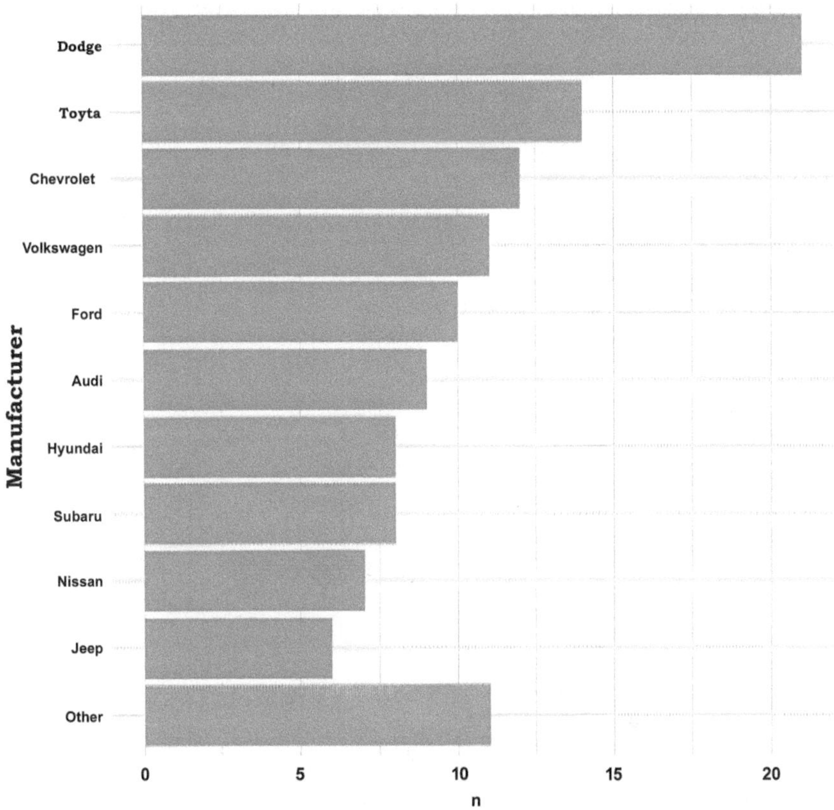

Figure 2.1 Bar plot.

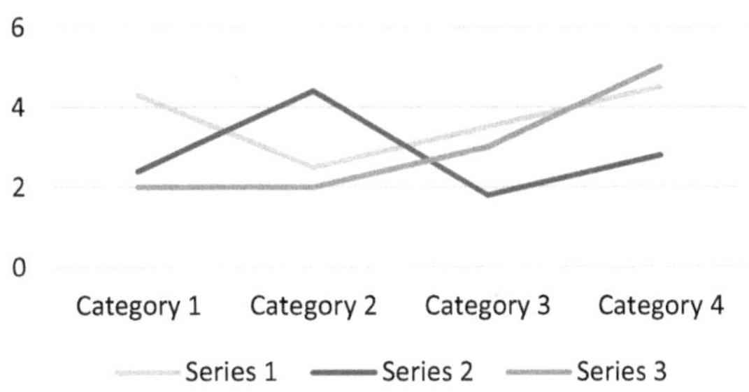

Figure 2.2 Line chart.

- **Details:** Scatter plots are used to observe the relationship between two continuous variables. By plotting one variable on the x-axis and another on the y-axis, researchers can identify trends, clusters, and potential outliers. This type of visualization is crucial in fields such as statistics, economics, and biology.

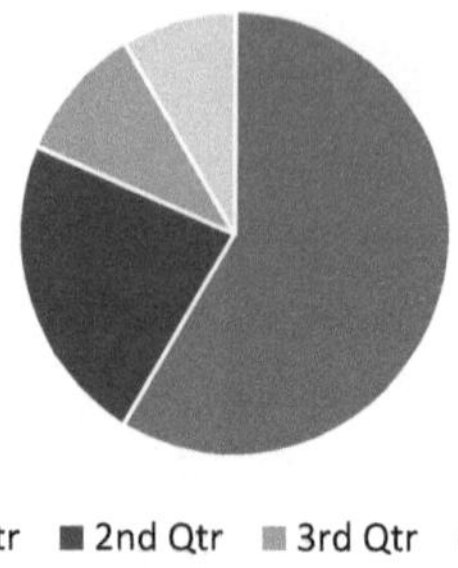

Figure 2.3 Pie chart.

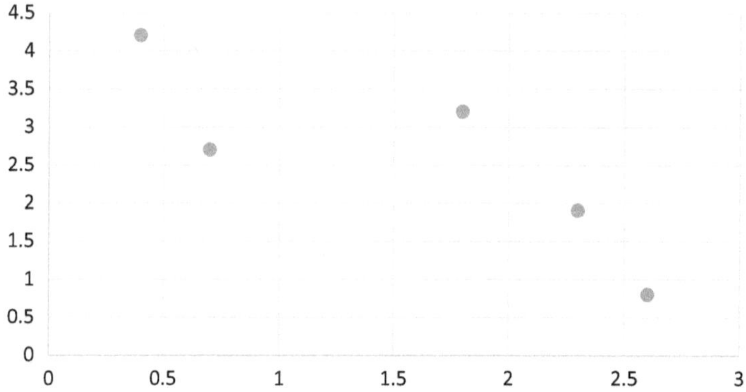

Figure 2.4 Scatter plots.

- Heat Maps (Guo et al., 2020):
  - **Purpose:** Heat maps are effective for showing data density and variations across different variables as represented in Figure 2.5. They use color gradients to represent the magnitude of values.
  - **Details:** Heat maps display complex datasets by using colors to indicate the value of data points. They are particularly useful in showing the intensity of variables across a matrix format, making it easier to spot trends and correlations. This type of visualization is often used in fields like bioinformatics, geography, and network analysis.

2.2.4 Tools and technologies

In today's data-driven world, effectively interpreting and communicating complex information is crucial. Data visualization tools play a pivotal role in transforming raw data into visually engaging and comprehensible insights, enabling users to make informed decisions (Islam and Jin, 2019). Among the myriad of available tools, Tableau, Power BI, D3.js, Matplotlib, and Seaborn stand out for their unique features and capabilities (Cossich

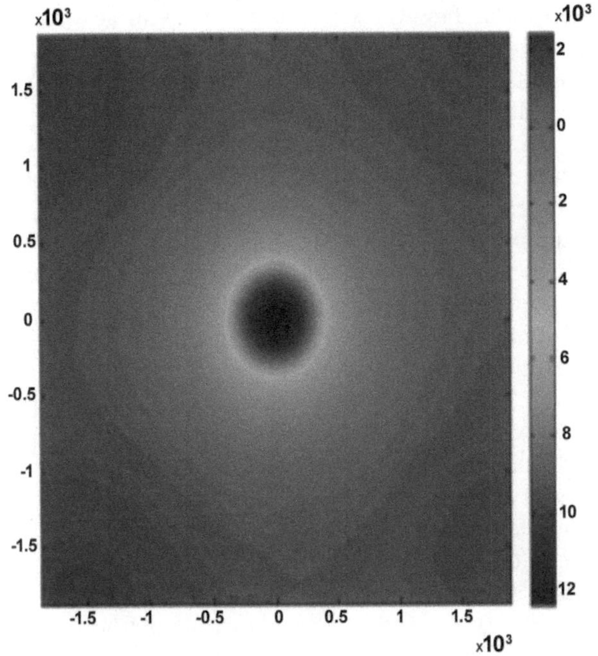

Figure 2.5 Heat map.

et al., 2023). Each tool offers distinct advantages tailored to different user needs, from intuitive interfaces for beginners to powerful customization options for advanced users. This section delves into these popular tools, highlighting their key features, strengths, and applications.

- **Tableau:**
  - Summary: Tableau is well-known for its intuitive interface and robust data visualization capabilities, enabling users to generate a diverse range of visualizations, from basic bar charts to intricate interactive dashboards.
  - Description: Tableau excels in its capacity to connect with various data sources, conduct real-time data analysis, and produce visually striking and informative graphics. Its drag-and-drop functionality makes it accessible to users with varying levels of technical proficiency. Widely utilized in business intelligence, Tableau swiftly transforms large datasets into actionable insights.
- **Power BI:**
  - Summary: Developed by Microsoft, Power BI offers comprehensive data integration and visualization features, aiming to transform raw data into coherent, visually captivating, and interactive insights.

- Description: Highly esteemed for its seamless integration with other Microsoft products such as Excel and Azure, Power BI is a formidable tool for enterprises entrenched in the Microsoft ecosystem. It supports a broad spectrum of data sources and provides tools for data preparation, discovery, and interactive dashboard creation. It is renowned for its ability to handle substantial datasets and extensive customization options.

- **D3.js: Summary:**
  - D3.js is a JavaScript library facilitating the creation of dynamic, interactive data visualizations within web browsers, leveraging web standards like SVG, HTML, and CSS.
  - Description: D3.js distinguishes itself through its flexibility and potency, empowering developers to bind data to the Document Object Model (DOM) and apply data-driven transformations to the document. Widely employed for crafting intricate and highly tailored visualizations that are interactive and animated, it necessitates a solid grasp of JavaScript and web technologies, catering more to developers and technically proficient users.
- **Matplotlib and Seaborn:**
  - Summary: Matplotlib and Seaborn are Python libraries tailored for crafting static, animated, and interactive visualizations.
  - Description:
    - Matplotlib: Renowned for its versatility, Matplotlib offers an extensive range of plotting functions, spanning from straightforward line plots to intricate 3D graphs. Highly customizable, it allows users to fine-tune nearly every aspect of the plots and serves as a foundational tool for various other Python plotting libraries.
    - Seaborn: Built atop Matplotlib, Seaborn simplifies the creation of appealing and informative statistical graphics. Furnishing a high-level interface for crafting attractive statistical graphics proves particularly beneficial for visualizing complex datasets. Leveraging built-in themes and color palettes, Seaborn streamlines the process of generating aesthetically pleasing and insightful visualizations.

2.3 DATA MINING: EXTRACTING KNOWLEDGE FROM DATA

2.3.1 Definition and scope

Definition: Data mining is the process of discovering patterns, correlations, anomalies, and statistically significant structures and knowledge from large datasets using various techniques from statistics, machine learning, and database systems (Gupta and Chandra, 2020). It involves the use of sophisticated algorithms to extract and identify useful information that would otherwise remain buried in the vast amounts of data collected by organizations.

Scope: The scope of data mining extends across multiple domains and industries due to its ability to handle diverse data sources and formats (Ikegwu et al., 2022). It encompasses various stages, including data collection, preprocessing, pattern discovery, and post-processing to validate and interpret the patterns.

1. **Data Sources:** Data mining can utilize multiple types of data sources, such as:
    - **Databases:** Structured collections of data that are typically organized into tables.
    - **Data Warehouses:** Large repositories that store data from multiple sources, often used for reporting and data analysis.
    - **Internet:** Vast amounts of unstructured data available online, including social media, blogs, and web pages.
    - **Other Large Repositories:** This includes sensor data, transactional data, and logs generated by various applications and devices.
2. **Techniques:** The process employs various techniques such as:
    - **Classification:** Assigning data to predefined categories or classes.
    - **Clustering:** Grouping data points into clusters based on similarity.
    - **Association Rule Learning:** Identifying interesting relationships between variables.
    - **Anomaly Detection:** Detecting outliers or abnormal data points that do not fit the general pattern.
    - **Regression:** Predicting a continuous value based on the relationships among variables.
    - **Summarization:** Providing a compact representation of the dataset.
3. **Applications:** Data mining has a wide range of applications, including but not limited to:
    - **Market Analysis and Management:** Predicting customer behavior, identifying potential customers, and market segmentation.
    - **Fraud Detection:** Identifying fraudulent activities in finance, insurance, and other sectors.
    - **Healthcare:** Diagnosing diseases, discovering patient patterns, and improving treatments.
    - **Manufacturing:** Quality control, process optimization, and predictive maintenance.
    - **Telecommunications:** Improving customer service, network optimization, and fraud detection.
4. **Challenges:** Despite its powerful capabilities, data mining faces several challenges, such as:
    - **Data Quality:** Incomplete, noisy, and inconsistent data can hinder the accuracy of mining results.
    - **Scalability:** Handling massive datasets efficiently requires scalable algorithms and computing power.

- **Privacy and Security:** Ensuring the privacy and security of sensitive information during the data mining process.
- **Interpretability:** Making the discovered patterns and models understandable to non-expert users.

Data mining thus represents a critical tool in the extraction of actionable insights from vast datasets, driving informed decision-making and innovation across various fields.

2.3.2 Key data mining techniques

Key techniques in data mining include:

- **Classification:** This technique involves assigning items to predefined categories or classes. It is widely used in scenarios where the goal is to predict categorical labels for new data points (Banerjee et al., 2023). Common algorithms for classification include decision trees, random forests, support vector machines (SVMs), and neural networks. For example, in email filtering, emails are classified as "spam" or "not spam" based on various features like the sender's address, the email's content, and the presence of certain keywords.
- **Clustering:** Clustering involves grouping a set of objects in such a way that objects in the same group (called a cluster) are more similar to each other than to those in other groups. This unsupervised learning technique is often used in exploratory data analysis to find natural groupings within data. Common clustering algorithms include K-Means, hierarchical clustering, and DBSCAN. Applications include customer segmentation in marketing and identifying areas of similar vegetation in satellite imagery.
- **Association Rule Mining:** This technique is used to discover interesting relations between variables in large databases (Telikani et al., 2020). It identifies rules that highlight the relationships between variables, such as "If a customer buys bread, they are likely to buy butter." These rules are typically represented in the form of if-then statements. Association rule mining is extensively used in market basket analysis, helping retailers understand purchase patterns.
- **Regression Analysis:** Regression analysis involves estimating the relationships among variables. It is primarily used for prediction and forecasting, where the objective is to predict a continuous outcome variable based on one or more predictor variables. Linear regression, logistic regression, and polynomial regression are common techniques. For instance, regression analysis can predict housing prices based on factors like location, size, and age of the property.
- **Anomaly Detection:** This technique focuses on identifying unusual data records that might be interesting or represent data errors. Anomalies can indicate critical incidents such as fraud, network

intrusions, or errors in data entry. Anomaly detection methods include statistical techniques, clustering-based methods, and machine learning approaches like isolation forests. It is widely used in fraud detection systems, network security, and quality control processes.

2.3.3 Data mining process (Banimustafa and Hardy, 2020)

1. **Data Cleaning:** This step involves removing noise and inconsistencies from the data. It includes handling missing values, correcting errors, and eliminating duplicates. Data cleaning is crucial as the quality of data directly impacts the accuracy of the data mining results.
2. **Data Integration:** In this step, data from multiple sources are combined to create a unified dataset. This process may involve schema integration, data consolidation, and resolving data conflicts. Data integration ensures that all relevant information is available in a single, coherent dataset for analysis.
3. **Data Selection:** During data selection, relevant data is chosen for the analysis. This involves selecting subsets of data that are necessary for the data mining task, based on certain criteria. Effective data selection helps focus the analysis on the most pertinent data, improving efficiency and relevance.
4. **Data Transformation:** This step converts data into appropriate forms for mining. It may involve normalization, aggregation, and generalization of data. Data transformation ensures that the data is in a suitable format and scale for the application of data mining algorithms.
5. **Data Mining:** This is the core step of the process, where algorithms are applied to extract patterns from the data. Various data mining techniques such as classification, clustering, association rule mining, and regression are employed based on the specific objectives of the analysis.
6. **Pattern Evaluation:** In this step, the patterns discovered during data mining are evaluated to identify the truly interesting ones that represent knowledge. This involves assessing the relevance, validity, and novelty of the patterns. Pattern evaluation helps in distinguishing significant insights from trivial findings.
7. **Knowledge Presentation:** The final step involves using visualization and knowledge representation techniques to present the mined knowledge to users. Effective presentation ensures that the insights are communicated clearly and can be easily interpreted and acted upon. This might involve creating charts, graphs, reports, or interactive dashboards.

2.3.4 Tools and technologies (Al-Hashedi and Magalingam, 2021)

- **RapidMiner:** RapidMiner is an open-source platform for data science that offers an integrated environment for data preparation, machine learning, deep learning, text mining, and predictive analytics. It is user-friendly and supports a wide range of data mining tasks, making it popular among both beginners and experts.

- **WEKA:** WEKA (Waikato Environment for Knowledge Analysis) is a collection of machine learning algorithms for data mining tasks. It provides tools for data preprocessing, classification, regression, clustering, association rules, and visualization. WEKA is well-regarded for its ease of use and comprehensive suite of algorithms.
- **Apache Mahout:** Apache Mahout is a library of scalable machine learning algorithms designed to work with big data. It includes implementations for clustering, classification, and collaborative filtering. Mahout is built to run on Hadoop, allowing it to process large datasets efficiently.
- **KNIME:** KNIME (Konstanz Information Miner) is an open-source data analytics, reporting, and integration platform. It supports a wide range of data mining and machine learning tasks through its modular data pipelining concept. KNIME is appreciated for its intuitive interface and ability to integrate with various other tools and libraries.

By understanding and utilizing these techniques, processes, and tools, organizations can effectively harness the power of data mining to uncover valuable insights and drive data-driven decision-making.

2.4 SYNERGY BETWEEN DATA VISUALIZATION AND DATA MINING

2.4.1 Complementary nature

Data visualization and data mining are two interconnected disciplines that, when combined, can significantly enhance the process of data analysis and interpretation. They complement each other in various ways, each enhancing the effectiveness of the other.

2.4.1.1 Data visualization for data mining (Romero and Ventura, 2020)

Understanding Data Distributions: Data visualization plays a crucial role in the preliminary stages of data mining by providing a clear and intuitive understanding of data distributions. Visual tools such as histograms, box plots, and scatter plots allow analysts to quickly grasp the underlying structure and distribution of the data. This visual insight can highlight patterns, trends, and anomalies that might not be immediately apparent in raw data.

- **Identifying Outliers:** Visual representations can easily show outliers or unusual data points that might indicate errors or significant phenomena requiring further investigation.

- **Detecting Patterns:** Visualization tools can help identify trends, clusters, or correlations within the data, guiding the choice of appropriate data mining techniques.
- **Assessing Data Quality:** Visual checks can reveal inconsistencies, missing values, or skewed distributions, prompting necessary data cleaning and preprocessing steps before mining.

Guiding the Mining Process: By visualizing data distributions, analysts can better determine where to focus their mining efforts. For instance, visualizations can help in identifying which variables are most relevant for a particular analysis, thus reducing the dimensionality and complexity of the data mining task.

- **Targeted Analysis:** Visual insights can direct attention to specific areas or segments of the data that are more likely to yield significant results.
- **Refining Hypotheses:** Visualization helps in formulating and refining hypotheses that can be tested through data mining techniques.

2.4.1.2 Data mining for data visualization (Romero and Ventura, 2020)

Making Mining Results Comprehensible: Once data mining techniques have been applied, the results often need to be communicated to stakeholders or used for decision-making. Here, data visualization comes into play by translating complex mining outputs into comprehensible visual formats.

- **Visual Representation of Patterns:** Techniques such as clustering and association rule mining can produce results that are complex to interpret in their raw form. Visual tools like dendrograms, heatmaps, and network diagrams make these patterns easily understandable.
- **Summarizing Results:** Visualization can summarize key findings from mining activities, such as frequent itemsets, classification outcomes, or regression trends, in an accessible and digestible manner.

Enhancing Interpretability and Actionability: Effective visual representation of mining results enhances their interpretability, making it easier for decision-makers to grasp the insights and take informed actions.

- **Interactive Dashboards:** Interactive visualizations allow users to explore mining results dynamically, enabling deeper insights and more nuanced understanding.
- **Storytelling with Data:** Visualizations can be used to tell compelling stories about the data, illustrating the impact of certain patterns and trends in a way that resonates with a broader audience.

Feedback Loop for Continuous Improvement: The interplay between data visualization and data mining can create a feedback loop that continuously improves the quality and depth of analysis.

- **Iterative Refinement:** Visualization of mining results can lead to new questions and hypotheses, prompting further rounds of data mining and analysis.
- **Data-Driven Decisions:** As visualized results are used to make decisions, new data generated from these decisions can be fed back into the mining process, enhancing the overall analytical framework.

By integrating data visualization and data mining, organizations can leverage the strengths of both approaches to achieve a more comprehensive and insightful analysis. Visualization aids in the initial exploration and understanding of data, guiding effective data mining, while the results of data mining can be effectively communicated and interpreted through visualization. This synergy not only enhances the clarity and impact of data-driven insights but also fosters a more iterative and dynamic analytical process.

2.5 CASE STUDIES (MOHD SELAMAT ET AL., 2020)

2.5.1 Retail industry

Uncovering Purchasing Patterns: In the retail industry, data mining techniques such as association rule learning, clustering, and classification can uncover intricate purchasing patterns and behaviors. For example, market basket analysis can reveal products frequently bought together, while clustering techniques can segment customers based on purchasing behavior, demographics, and other relevant features.

Visualizing Insights: Visualization tools then come into play to present these insights in an accessible and actionable format. Seasonal trends, for instance, can be visualized using time-series plots that display sales fluctuations throughout the year. Customer preferences can be illustrated through heatmaps showing the popularity of products across different regions or customer segments.

Case Example: Consider a large retail chain that uses data mining to analyze transaction data. By applying association rule mining, the chain discovers that customers who purchase baby diapers are also likely to buy baby wipes and formula. This finding can be visualized through network diagrams showing the strength of associations between different products. Additionally, time-series analysis might reveal that these purchases spike during certain times of the year, such as during holiday seasons. These visualizations enable the retail chain to optimize inventory, tailor marketing strategies, and enhance customer satisfaction.

2.5.2 Healthcare

Identifying Patterns in Patient Data: In the healthcare sector, data mining techniques like clustering, classification, and anomaly detection can identify patterns and trends within patient data. For example, clustering can group patients based on similar symptoms or treatment responses, while classification can predict disease outcomes based on historical data.

Informing Clinical Decisions: Visualizing these patterns helps clinicians and healthcare administrators make informed decisions. For instance, patient clusters can be visualized using scatter plots or cluster heatmaps, highlighting common characteristics within each group. Survival analysis curves can visualize the effectiveness of different treatment plans, aiding in selecting the best course of action for patient care.

Case Example: A hospital system uses data mining to analyze electronic health records (EHRs). By applying clustering algorithms, they identify groups of patients with similar risk profiles for diabetes. These clusters are visualized through radar charts and scatter plots, showing the distribution of risk factors such as age, BMI, and family history. This visualization helps healthcare providers design targeted intervention programs. Additionally, classification models predict the likelihood of hospital readmissions, and the results are visualized through probability heatmaps, assisting in allocating resources more effectively to high-risk patients.

2.6 CHALLENGES AND SOLUTIONS

2.6.1 Data quality

Challenge: Poor data quality can lead to misleading visualizations and mining results. Inaccurate, incomplete, or inconsistent data can skew analysis, resulting in erroneous conclusions that can misguide decision-making processes.

Solution: Implement robust data cleaning and preprocessing steps to ensure high-quality input data. This involves several key actions:

- **Data Cleaning:** Identifying and correcting errors, such as duplicates, outliers, and inconsistencies.
- **Data Imputation:** Filling in missing values using techniques like mean/mode imputation, k-nearest neighbors, or more advanced statistical methods.
- **Normalization and Standardization:** Ensuring data is in a consistent format and scale, which is crucial for accurate analysis.
- **Validation and Verification:** Continuously checking data for accuracy and reliability through cross-validation and other validation techniques.

Case Example: A financial services company handles large volumes of transaction data daily. To maintain high data quality, they employ automated scripts for data cleaning that remove duplicates and correct errors. They also use advanced imputation techniques to handle missing data and perform regular audits to validate data accuracy. These preprocessing steps ensure that the subsequent data mining and visualization efforts yield reliable insights, enhancing the company's risk management and decision-making processes.

2.6.2 Scalability

Challenge: Handling large datasets efficiently. As data volumes grow, traditional data processing techniques and tools may struggle to keep up, leading to performance bottlenecks and slow analysis times.

Solution: Utilize scalable algorithms and distributed computing frameworks such as Apache Hadoop and Spark. These frameworks enable the processing of large datasets across distributed computing resources, significantly improving efficiency and speed.

- **Apache Hadoop:** A framework that allows for the distributed processing of large datasets across clusters of computers using simple programming models. It's highly scalable and can handle petabytes of data.
- **Apache Spark:** An open-source distributed computing system that provides an interface for programming entire clusters with implicit data parallelism and fault tolerance. Spark is known for its speed and ease of use, particularly in big data processing tasks.

Case Example: An e-commerce platform with millions of daily transactions adopts Apache Spark to handle its data processing needs. By distributing the computation across a cluster, Spark processes massive datasets quickly and efficiently, enabling real-time analytics and insights. This scalability allows the platform to maintain high performance even as data volumes grow, supporting better customer experiences and more agile business operations.

2.6.3 Interpretability

Challenge: Complex models and visualizations can be difficult to interpret. Advanced data mining models such as deep learning networks, while powerful, often operate as "black boxes," making it challenging to understand how they arrive at their conclusions. Similarly, overly complex visualizations can confuse rather than clarify.

Solution: Focus on creating intuitive visualizations and using interpretable models or providing clear explanations of complex models.

- **Simplified Models:** Where possible, use simpler, more interpretable models such as decision trees or linear regression. When more complex models are necessary, use techniques like SHAP (SHapley Additive

explanations) values or LIME (Local Interpretable Model-agnostic Explanations) to interpret the results.
- **Clear Visualizations:** Design visualizations that are straightforward and avoid unnecessary complexity. Use clear labels, legends, and annotations to aid understanding. Interactive visualizations can also help users explore data more intuitively.

Case Example: A healthcare analytics firm uses a deep learning model to predict patient outcomes. To make the model's predictions interpretable, they employ SHAP values to explain the contribution of each feature to the predictions. The results are then visualized using bar charts and summary plots that clearly indicate the impact of different variables on patient outcomes. Additionally, the firm creates interactive dashboards that allow clinicians to explore the data and model outputs dynamically, enhancing their understanding and trust in the predictive insights.

By addressing these challenges with targeted solutions, the integration of data visualization and data mining can be optimized, leading to more effective and insightful data analysis practices.

2.7 PRACTICAL APPLICATIONS (WANG ET AL., 2020)

2.7.1 Business intelligence

Customer Behavior Analysis: Businesses harness data visualization and data mining to gain profound insights into customer behavior. By analyzing transaction histories, clickstream data, and customer feedback, companies can segment customers based on their purchasing patterns, preferences, and demographics. Data mining techniques such as clustering and classification help identify distinct customer segments, while visualization tools like heatmaps and bubble charts illustrate these segments vividly.

- **Personalized Marketing:** Visualized insights from data mining enable businesses to create targeted marketing campaigns. For example, a retailer might use these insights to tailor promotions to specific customer groups, enhancing engagement and boosting sales.
- **Customer Churn Analysis:** Predictive modeling techniques, visualized through survival curves or risk heatmaps, can identify customers at risk of churn. Businesses can then develop retention strategies to mitigate this risk.

Sales Trend Analysis: Data visualization and mining reveal critical sales trends and patterns. Time-series analysis and visualizations like line charts and bar graphs illustrate sales performance over time, highlighting peaks and troughs.

- **Seasonal Trends:** Businesses can identify seasonal variations in sales and prepare accordingly. For instance, a retail chain might use historical sales data visualized through seasonal decomposition plots to stock inventory optimally for peak seasons.
- **Product Performance:** Sales data can be mined to determine top-performing products. Visualization tools can then display this information in an accessible format, such as Pareto charts, allowing businesses to focus on high-impact products.

Operational Efficiency: Data mining and visualization also play a pivotal role in optimizing operational efficiency. Process mining techniques can uncover inefficiencies in business processes, and these can be visualized using process maps and flowcharts.

- **Supply Chain Optimization:** Visualization of data from supply chain operations, such as delivery times and inventory levels, can identify bottlenecks and areas for improvement. Dashboards displaying key performance indicators (KPIs) help managers monitor and optimize operations in real time.

2.7.2 Scientific research

Experimental Data Analysis: Researchers leverage data mining and visualization to analyze complex experimental data. Techniques such as regression analysis, clustering, and anomaly detection help in uncovering significant patterns and relationships within the data.

- **Hypothesis Testing:** Visual tools like scatter plots, histograms, and box plots enable researchers to visualize distributions, correlations, and trends, aiding in hypothesis testing and validation.
- **Genomic Research:** In genomics, data mining techniques identify gene expression patterns, and these are visualized through heatmaps, volcano plots, and network diagrams, providing insights into gene functions and interactions.

Correlation and Causation: Understanding correlations and potential causations is crucial in scientific research. Data mining can reveal these relationships, while visualization makes them comprehensible.

- **Network Analysis:** Researchers use network diagrams to visualize relationships between variables, such as protein-protein interactions or social network connections in epidemiology studies.
- **Time-Series Analysis:** Longitudinal studies often involve time-series data. Visualization techniques like time-series plots and autocorrelation plots help in analyzing temporal patterns and trends.

Advancing Knowledge: Effective visualization of research findings facilitates knowledge dissemination and comprehension.

- **Scientific Publications:** Graphs, charts, and other visual aids are integral to scientific publications, making complex data and results accessible to a broader audience.
- **Interactive Data Exploration:** Tools like Jupyter Notebook enable interactive data exploration and visualization, allowing researchers to share their workflows and results dynamically.

2.7.3 Social media analysis

Public Sentiment Analysis: Mining social media data helps in understanding public sentiment toward brands, products, or events. Sentiment analysis techniques, combined with natural language processing (NLP), extract opinions and sentiments from social media posts.

- **Sentiment Visualization:** Word clouds, sentiment heatmaps, and time-series plots showing sentiment trends over time help businesses gauge public opinion and react promptly.

Tracking Brand Performance: Social media interactions provide valuable insights into brand performance. Data mining techniques analyze these interactions to uncover engagement patterns, influencer impact, and campaign effectiveness.

- **Engagement Metrics:** Visualizing metrics such as likes, shares, comments, and mentions through dashboards helps companies monitor their social media performance in real time.
- **Influencer Analysis:** Network diagrams can illustrate influencer relationships and their impact on brand perception, helping companies identify and collaborate with key influencers.

Tailoring Marketing Strategies: By understanding trends and patterns in social media data, companies can tailor their marketing strategies to better align with audience preferences and behaviors.

- **Trend Analysis:** Data mining reveals trending topics and hashtags, which can be visualized using trend graphs and topic clusters, allowing companies to align their content with current trends.
- **Campaign Optimization:** Visualization of engagement data helps in assessing the effectiveness of marketing campaigns and refining strategies for better results.

2.8 FRAUD DETECTION

Detecting Unusual Patterns: Data mining techniques such as anomaly detection, clustering, and classification are essential in identifying fraudulent activities. These techniques analyze large volumes of transactional data to spot unusual patterns and behaviors indicative of fraud.

- **Anomaly Detection:** Algorithms identify outliers in data, which can be visualized through scatter plots, box plots, and anomaly heatmaps, highlighting potential fraud cases.
- **Pattern Recognition:** Clustering techniques group similar fraudulent transactions, and these groups can be visualized using dendrograms and cluster maps to understand common fraud patterns.

Visualizing Fraudulent Activities: Visualization tools transform complex data mining results into comprehensible formats, aiding analysts in investigating and responding to fraud.

- **Fraud Network Diagrams:** Network diagrams can illustrate connections between fraudulent activities, revealing the structure of fraud rings and helping in their dismantling.
- **Heatmaps:** Transaction heatmaps can show regions or time periods with high concentrations of suspicious activities, directing analysts' focus to high-risk areas.

Prompt Response and Prevention: Effective visualization enables quick detection and response to fraud, minimizing potential damage.

- **Real-Time Dashboards:** Interactive dashboards provide real-time monitoring of transactions, alerting analysts to suspicious activities immediately. Visualization tools such as real-time transaction maps and anomaly alerts facilitate swift action.
- **Predictive Analytics:** By visualizing predictive models, businesses can proactively prevent fraud. For example, predictive risk scores visualized through risk dashboards can help financial institutions preemptively block high-risk transactions.

2.9 FUTURE TRENDS AND INNOVATIONS

2.9.1 AI and machine learning integration

The integration of AI and machine learning with data visualization and mining is revolutionizing the way complex datasets are processed and analyzed. These advanced technologies enable more sophisticated pattern recognition, predictive analytics, and automated data processing, significantly enhancing the speed and accuracy of insights derived from vast and intricate data. Machine learning algorithms can identify hidden patterns and trends that might be missed by traditional methods, while AI-driven visualizations can adapt and update in real time, providing dynamic and interactive representations of data. This synergy between AI, machine learning, and data analytics is paving the way for more intelligent and autonomous decision-making processes across various industries.

2.9.2 Real-time data analysis

The increasing demand for real-time insights is propelling the development of technologies capable of visualizing and mining data instantaneously. Real-time data analysis allows organizations to monitor and respond to changes as they happen, providing a significant competitive advantage in fast-paced environments. Technologies like streaming analytics platforms, real-time dashboards, and in-memory computing are being utilized to process and visualize data without delay. This capability is particularly critical in sectors such as finance, healthcare, and e-commerce, where timely data-driven decisions can lead to improved outcomes, enhanced customer experiences, and greater operational efficiency.

2.9.3 Advanced visualization techniques

Emerging techniques like virtual reality (VR) and augmented reality (AR) are being explored to create more immersive and interactive visualizations. These advanced visualization methods offer a three-dimensional perspective on data, allowing users to interact with complex datasets in a more intuitive and engaging manner. For example, VR can provide a fully immersive environment where users can navigate through data landscapes, while AR can overlay data visualizations onto the real world, enhancing context and comprehension. These innovations are poised to transform fields such as scientific research, education, and business analytics by providing deeper insights and more impactful data presentations.

2.9.4 Ethical considerations

As data analysis capabilities expand, so do the ethical and privacy concerns associated with the use of data. It is imperative to develop robust frameworks and guidelines that ensure the responsible and ethical use of data. This includes implementing strict data governance policies, ensuring transparency in data collection and analysis processes, and protecting individuals' privacy rights. Ethical considerations also encompass the mitigation of biases in data mining and machine learning models to ensure fair and equitable outcomes. As organizations increasingly rely on data-driven decision-making, prioritizing ethical standards is crucial to maintaining public trust and fostering a sustainable and responsible data ecosystem.

2.10 CONCLUSION

Data visualization and data mining have become indispensable tools in today's data-driven world, transforming raw, unstructured data into meaningful, actionable insights that drive innovation and efficiency across various sectors. These tools enhance decision-making by visually representing complex datasets, allowing decision-makers to identify trends, correlations, and anomalies that would be difficult to discern from raw data alone. This leads to more accurate forecasting, better strategic planning, and improved operational efficiency. By automating the data analysis process, data mining reduces the time and effort required to uncover important patterns, while visualization tools simplify the interpretation of these patterns, making it easier for stakeholders to understand and act upon the insights generated. This combination streamlines workflows and enhances productivity. Organizations that effectively utilize these tools are better positioned to innovate and maintain a competitive edge, quickly adapting to market changes, identifying new opportunities, and optimizing operations. The versatility of data visualization and mining techniques makes them applicable across various fields, from scientific research and social media analysis to fraud detection and healthcare. To fully harness their potential, organizations must ensure data quality, employ scalable solutions, focus on interpretability, embrace emerging technologies like VR and AR, and address ethical considerations. By following these practices and leveraging technological advancements, organizations can unlock the full potential of data visualization and mining, enhancing their ability to make data-driven decisions and uncover hidden opportunities in the modern data landscape. This chapter provides a comprehensive overview of data visualization and data mining, highlighting their principles, techniques, and applications. By understanding and applying these concepts, one can unlock the transformative power of data.

REFERENCES

Al-Hashedi, K. G., & Magalingam, P. (2021). Financial fraud detection applying data mining techniques: A comprehensive review from 2009 to 2019. *Computer Science Review*, *40*, 100402.

Allil, K. (2024). Integrating AI-driven marketing analytics techniques into the classroom: pedagogical strategies for enhancing student engagement and future business success. *Journal of Marketing Analytics*, 12, 1–27.

Banerjee, P., Chattopadhyay, T., & Chattopadhyay, A. K. (2023). Comparison among different Clustering and Classification Techniques: Astronomical data-dependent study. *New Astronomy*, *100*, 101973.

Banimustafa, A., & Hardy, N. (2020). A scientific knowledge discovery and data mining process model for metabolomics. *IEEE Access*, *8*, 209964–210005.

Cossich, V. R., Carlgren, D., Holash, R. J., & Katz, L. (2023). Technological breakthroughs in sport: current practice and future potential of artificial intelligence, virtual reality, augmented reality, and modern data visualization in performance analysis. *Applied Sciences*, *13*(23), 12965.

Guo, H., Zhang, W., Ni, C., Cai, Z., Chen, S., & Huang, X. (2020). Heat map visualization for electrocardiogram data analysis. *BMC Cardiovascular Disorders*, *20*, 1–8.

Gupta, M. K., & Chandra, P. (2020). A comprehensive survey of data mining. *International Journal of Information Technology*, *12*(4), 1243–1257.

Ikegwu, A. C., Nweke, H. F., Anikwe, C. V., Alo, U. R., & Okonkwo, O. R. (2022). Big data analytics for data-driven industry: A review of data sources, tools, challenges, solutions, and research directions. *Cluster Computing*, *25*(5), 3343–3387.

Islam, M., & Jin, S. (2019, November). An overview of data visualization. In *2019 International Conference on Information Science and Communications Technologies (ICISCT)* (pp. 1–7). IEEE.

Keim, D. A., Hao, M. C., Dayal, U., Janetzko, H., & Bak, P. (2010). Generalized scatter plots. *Information Visualization*, *9*(4), 301–311.

Lal, J., Mitkari, A., Bhosale, M., & Doermann, D. (2023, August). Lineformer: Line chart data extraction using instance segmentation. In *International Conference on Document Analysis and Recognition* (pp. 387–400). Springer Nature Switzerland.

Luo, X., Yuan, Y., Zhang, K., Xia, J., Zhou, Z., Chang, L., & Gu, T. (2019). Enhancing statistical charts: toward better data visualization and analysis. *Journal of Visualization*, *22*, 819–832

Midway, S. R. (2020). Principles of effective data visualization. *Patterns*, *1*(9).

Mohd Selamat, S. A., Prakoonwit, S., & Khan, W. (2020). A review of data mining in knowledge management: applications/findings for transportation of small and medium enterprises. *SN Applied Sciences*, *2*(5), 818.

Myatt, G. J., & Johnson, W. P. (2009). Making sense of data II: A practical guide to data visualization, advanced data mining methods, and applications (Vol. 2). John Wiley & Sons.

Qin, X., Luo, Y., Tang, N., & Li, G. (2020). Making data visualization more efficient and effective: a survey. *The VLDB Journal*, *29*(1), 93–117.

Romero, C., & Ventura, S. (2020). Educational data mining and learning analytics: An updated survey. *Wiley Interdisciplinary Reviews: Data mining and Knowledge Discovery*, *10*(3), e1355.

Shu, X., & Ye, Y. (2023). Knowledge Discovery: Methods from data mining and machine learning. *Social Science Research*, *110*, 102817.

Siirtola, H. (2019, July). The cost of pie charts. In *2019 23rd International Conference Information Visualisation (IV)* (pp. 151–156). IEEE.

Talbot, J., Setlur, V., & Anand, A. (2014). Four experiments on the perception of bar charts. *IEEE Transactions on Visualization and Computer Graphics*, *20*(12), 2152–2160.

Telikani, A., Gandomi, A. H., & Shahbahrami, A. (2020). A survey of evolutionary computation for association rule mining. *Information Sciences*, *524*, 318–352.

Wang, D., Miwa, T., & Morikawa, T. (2020). Big trajectory data mining: A survey of methods, applications, and services. *Sensors*, *20*(16), 4571.

Chapter 3

Leveraging data-driven design for consumer-centric robust product development

Debraj Bhattacharjee and Jayshree Patnaik

3.1 INTRODUCTION

The creation of new products is an essential part of technological advancement that continuously modifies consumers' lifestyles and drastically improves the level of living (Williams, 2007). Most inventions and items produced, distributed, and developed within the global economy are targeted towards consumers. Design now refers to any function that involves the design of a product or service, whether it be tangible or intangible (Holmlid, 2007). Originally, design referred to roles held largely by makers and craftsmen. The method of organising, generating, and executing ideas to enhance the natural setting has historically been regarded as design, with the fundamental aim of design as "the conception and realisation of new things". Today, design is bringing along the consumer product interface to a level, where the product adds value, improves, and enhances the human way of life (Veryzer & Borja de Mozota, 2005). Therefore, it is logical to pay additional attention to the product with the human interface when building things that are anticipated to be utilised by humans. Such an interface, when not properly built, may have negative effects. Furthermore, it has been widely acknowledged that during a design process, encompassing heterogeneity of user's preferences is a difficult task, and mandates a certain mindset and approach. It is quite a challenge to maintain such diversity throughout the concept development to the final product stage. The main objective should be to find and address any humanist consequences that may apply to similar usability during the conceptual design phase, where strong partnership and creativity offer a further view of the concept, design, and expected performance (Andreasen et al., 2015).

The vast majority of innovations today do not succeed due to insufficient or undesirable work at the initial stages of the design process. Weak customer research is the main reason: a dearth of precision in finding practical underserved customer demands and testing solution by actual user testing. Businesses are constantly seeking new ways to increase their consumer base. The fundamental goal of their survival is to increase earnings while providing a high-quality product; therefore, businesses are constantly seeking for new ways to

 DOI: 10.1201/9781003621102-3

reach a greater number of customers throughout the world and become the leading source (Codini et al., 2023). Discovering how a customer behaves and what products they value in their lives can have a significant influence on how well a product is perceived by the consumer. When a designer or innovator combines design thinking and inclusive design into the development or redesign of a product, they are able to strategically plan for inclusiveness into the design process. Understanding the users and their experiences necessitates research and careful decision-making. In order to incorporate inclusivity into design, designers and engineers must first understand the users' behaviours, attitudes, and the tasks that will be performed while using the product.

Dunne and Martin (2006) contrast design from design thinking, stating design thinking as mental methods used by designers rather than the designed products they create. Thus, Design is evolving beyond illustration, product, and interactive domain knowledge into a strategy for solving multifaceted problems in organisations and society as a result of this application. Moreover, design is evolving beyond illustration, product, and interactive domain knowledge into a strategy for solving multifaceted problems in organisations and society as a result of design thinking. Design thinking emerged in the twentieth century and has been implemented in a variety of sectors, including company management and new product development (Cross, 2023). Thus, design thinking is an approach that involves creating design concepts and applying them to a product/service/systems. Tim Brown popularises the concept of "design thinking" in business management by leveraging integrative thinking and the ability to blend customer feedback with the design of business organisations and team. A novel and innovative solution is produced via design thinking, and the model incorporates components from several methods to produce the best one.

Remarkable improvement in the product and service design can be achieved by considering the diverse customers' requirements and user-friendliness. Still, companies are failing to recognise or acknowledge the user requirement at the initial phase of product design process. Innovators must expand their considerations beyond technical aspects to effectively address the novel challenges arising from technologically driven product development in the twenty-first century, exemplified by the demands posed by artificial intelligence (AI) products and data-driven technologies. The capacity to assemble a brilliant team of experts from multiple disciplines, including management, engineering, accounting, and marketing, is the first step in the product design and development process.

3.2 PRODUCT DESIGN AND DEVELOPMENT PROCESS

The product design and development process according to Ulrich et al. (2008) can be defined with a six-phase process, namely, planning, concept development, system-level design, detail design, testing and refinement, and production ramp up. The six phases of product design and development produce lots of data that can be used to improve the product itself along with the product development process. Generally, two main areas where most products fail are due to improper market positioning and due to failure of the product while solving the pain point of the consumer (Branstad & Solem, 2020). The process of the product design starts with the identification of consumer pain points, which is generally part of the planning phase. This phase generally tries to find the market opportunity and defines the market segment. Parallelly, this phase also motivates the developers to explore the available products or develop new technology as per the requirement of the consumer. The next phase tries to develop multiple-concept product ideas by identifying the consumers' needs and finalising the concept product design through feasibility analysis. These multiple-concept designs are often called prototypes. The next phase of system-level design generates the product architectures and defines major subsystems to obtain the refined industrial design. Parallelly, the planning for product family extension and planning for target sales are done in this phase. The detail design phase is nothing but defining the product specification and finalising the marketing plan. Testing and refinement step is very significant as this helps the product designers to eliminate any major design-level flaws and for making the product more acceptable to the consumers. The refinement in the product design also helps in providing useful real-life testing results to the developer team. Production ramp up is the time to launch the product and sell the product to the targeted customers. This step continues in cycles for improving or developing new products or services for longer life in the market. As earlier mentioned, two main aspects of product design, namely, marketing and designing, play a critical role in the success of a product or service; different activities related to product design and development are provided in Table 3.1. These two aspects are further elaborated for the reader's understanding in Sections 3.3 and 3.4.

Table 3.1 Marketing and design aspects for product design and development process

Phases	*Marketing*	*Designing*
Planning phase	Identification of market opportunity and define market segment	Consider product architecture and the concept of new technologies
Concept development phase	Collect data for identification of customer needs. Finding the consumer base and research about the competitive products	Feasibility analysis of product concepts for confirming the industrial design concept. Testing of product concepts through lab experiments.
System-level design	Strech the product family and set the target sales price point	Generate alternative product architectures. Defining major subsystems and refinement of industrial design.

(Continued)

Table 3.1 (Continued)

Phases	*Marketing*	*Designing*
Detail design	Develop marketing plan	Define geometry, choose materials, and assign tolerance. Documentation of industrial design.
Testing and refinement	Promotion and launch materials enable field testing	Testing life, reliability, performance, regulatory and implement final design changes
Production ramp up	Early production	Testing the early production output

3.3 CONSUMER PREFERENCE IN PRODUCT DESIGN AND DEVELOPMENT

Consumers are the centre point of any product design process and their pain points are the basic building blocks for a successful product design. So, the identification of consumer preferences is a necessary step before starting the product design process. Considering this point, another important method for consumer problem solving is design thinking. Design thinking mainly focuses on careful observation of users' problems and providing a reliable solution, which can be provided in the form of a product. The design thinking process comprises steps such as empathise, define, ideate, prototype, and test as depicted in Figure 3.1 (Cross, 2023). The design thinking process tries to find the consumer pain point by careful observation and collection of full information about the needs to find the solution. The next phase is related

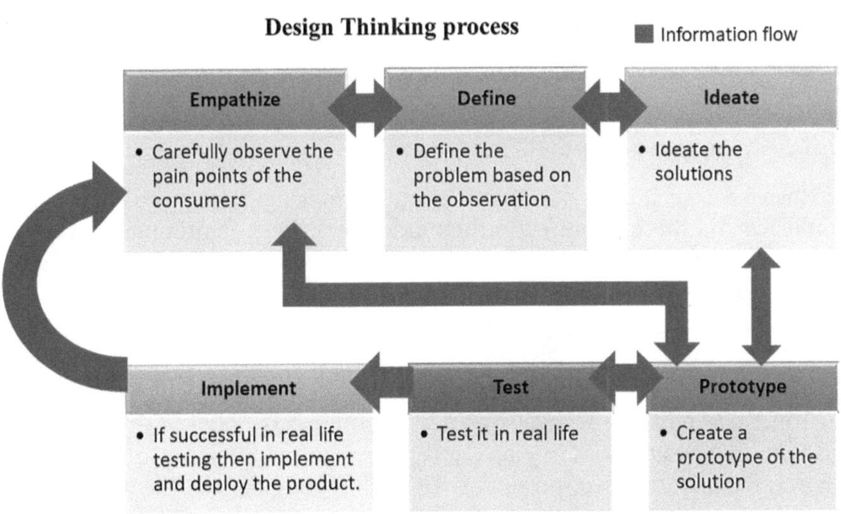

Figure 3.1 Design thinking process in product design and development.

to the process of defining the problem based on the collected information. The ideate phase is to list down the possible solutions for the defined problem. Next, the prototype phase builds the prototype of the possible solutions and tests for feasibility by collecting real-life test results. Finally, the best solution is considered for implementation and is deployed as a product or service.

So, the customer or consumer plays a big role in such methods, and the customer engagement is very important while developing the product according to design thinking. In case of new product development or existing product improvement, customer engagement has four inclusive design aspects, which are customer engagement and contribution, growing a large customer base, innovation and differentiation, and avoid retrofitting (Alston & DeKerchove, 2023).

3.3.1 Customer engagement and contribution

The ease of use of a product naturally increases customer engagement. So, if an existing product does not match the customer's preference or need, irregular engagement can be observed. This irregular engagement data can be used to find patterns, which may help the product designer identify the reasons behind the dissatisfaction, leading to possible improvement in product design.

3.3.2 Growing a larger customer base

It is not always true that if one set of consumers likes a product, it means all available customers should like it. So, a large customer base helps the designers to define the constraints in a concrete manner. This larger consumer base behaviour information also helps the product developer team to connect with a wider target audience by understanding the different usage contexts instead of the development team's ability.

3.3.3 Innovation and differentiation

Inclusive design can produce innovation by catering to the needs of a wider audience and this is required in the modern era for generating new product ideas. Also, the change in perspective and context can lead to new patterns, which can be used to identify the purpose of the solution.

3.3.4 Avoid retrofitting

In some companies, the inclusive design concept is only considered as an add-on. As a result of these, the concept of inclusive design is often considered at the final stages of product design. Now for example, the design solution for the disabled is often considered as an extra feature of the product.

3.4 DATA-DRIVEN PRODUCT DESIGN AND DEVELOPMENT

Data is a fuel for modern business models, and in the product development process, the data usage has increased in recent times. In the case of new product development and product improvement, consumer feedback is very important, and the importance of consumer feedback based on types of product users has been provided in Figure 11.2. The early adopters of any product are the first users of any product, and the feedback from them is very much important as they are the set of customers who will be experiencing the product (Nguyen et al., 2020). So, from an improvement perspective, these customers are a useful source of feedback about the product. Now the next set of customers are those who generally use the product after hearing about it from the early adopters. So, they already know the problems with the existing product. So, they are the source of the information about the minor problems with the product. The last two sets of users are the Late majority and Laggards, who are using the product once it is refined based on the feedback of the early adopters. Now, one may think that the feedback from the latter two groups is not that much important, but they are the users who provide the information about the pain points, which are not dealt with in the current form of the product. So, these users' feedback is important for developing a new product family, which may solve the problem of the prior version.

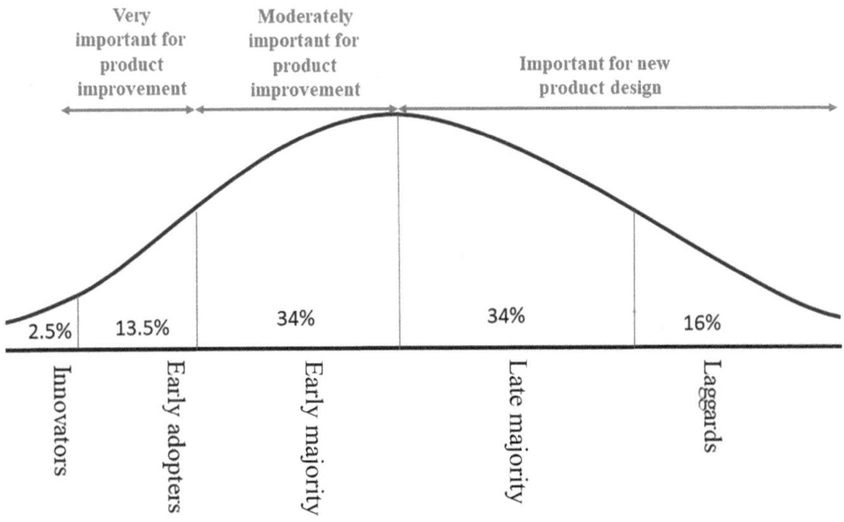

Figure 3.2 Importance of information according to the users (Sinek, 2011).

Based on the different levels of data usage, the organisations are categorised into three categories, namely, data deliberator, data adapter, and data innovator. Data deliberators are the organisations who are at the early phase of data strategy. The data adopters are the organisations, who use the data for business operations but still have scope for improvement. The data innovators are the organisations who are building data-driven strategies and using them at an advanced level to drive innovation in every field of business operations. According to the available data, the percentage of different categories is provided in Table 3.2.

Now, the consumer data can be observed using different sources in this age of modern information technology. The trend in social media, product review data, warranty claims data, guarantee claims data, product usage data, and consumer feedback are the types of data used for analysis to create useful insight (Fenko & van Rompay, 2018). Several tools are being used for improving the product design or developing a new product. In published literature and in practice, both qualitative and quantitative methods are being used for the product design and development process.

The data-driven product design and development process starts with a data collection method. The selection of the data source is the main concern when data-driven decisions need to be taken. Next is the consideration of different product design aspects while collecting different types of data. The different data sources need to be integrated for a holistic view of the product design and development process (Jreissat et al., 2017). This also indicates the importance of integration of internally generated data and externally generated data irrespective of the quantitative types or qualitative types. So, along with the collection of the data, storage is also important. The next part of the process is analysis of the data. Several data analysis methods can be used for such analysis. Some of the quantitative methods are discussed in

Table 3.2 Sector-wise data usage

Technology sector	*Data deliberations (%)*	*Data adaptors (%)*	*Data innovators (%)*
Technology	34	45	21
Financial	37	48	15
Communication and media	50	42	9
Manufacturing and media	64	28	8
Retail	52	42	6
Healthcare and life science	63	30	6
Public sector	60	37	3
Higher education	64	35	1

Source: (Bean, 2023).

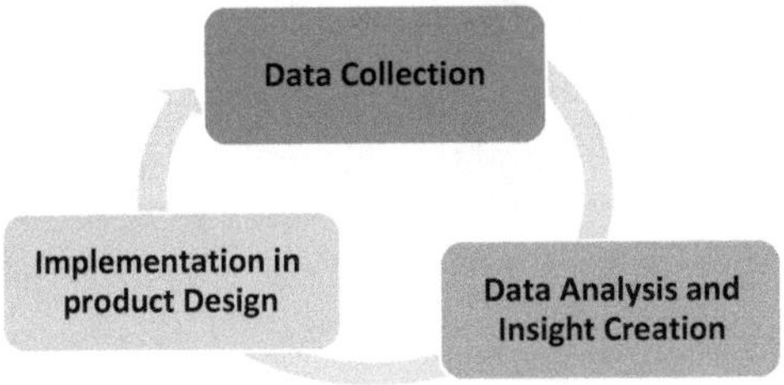

Figure 3.3 Data-driven product design.

this subsection. So, for such data analysis, a good data analysis team is required or a collaboration with a data analytics team is mandatory. Finally, after getting the insights, the implementation is required on the ground. This cycle is a continuous process and is required for a dynamic lean product design and development process. This eventually reduces the wastage of resources and reduces the failure chance.

Different data-driven approaches are used to capture consumer preferences and trends. These methods are presented in the following subsections.

3.4.1 Traditional methods

The traditional method like "House of Quality" is a good tool for product planning. The questions related to "What" are first asked, and the answers are collected from the customers and "How" are collected to find the available ways to solve the what using the available how's. Finally, by mapping these two pieces of information, the product planning matrix is developed.

3.4.2 Feature selection

Feature selection, in machine learning, is an approach which identifies the important features or product attributes in the case of product design and development. Also, based on consumer feedback, the manufacturers can concentrate on selected features or attributes to improve the product. The feature selection generally checks the information gained or the association between the features to find the set of best features.

3.4.3 Sentiment analysis and opinion mining

Sentiment analysis is a classification problem based on the features the sentiment is predicted. This sentiment can be defined either as positive or negative. This same concept can be used to predict the success of a product. The concept product can be tested before developing it. So, this may help in the reduction of product failure.

3.4.4 Segmentation or clustering

Clustering or segmentation is a process of grouping as per the similarity. This method is very useful to find the segment of customers, and further, the customer behaviour can be analysed for finding useful information, either using feature selection or sentiment analysis to find a better product idea or recommendation for change in the product design.

3.5 EXAMPLES OF DATA-DRIVEN CONSUMER-CENTRIC PRODUCT DESIGN AND DEVELOPMENT

3.5.1 Ubisoft data-driven game development

Ubisoft is an example where the entire product is developed based on user data (Werder et al., 2020). Ubisoft is a well-known game developer company and known for its fantastic games, namely, Assassin's Creed series, Far Cry series, Just Dance series, and Prince of Persia series. This company is known for its innovative and attractive game design, which are nothing but the result of good data-driven product development. For example, "Ghost Recon Wildlands" is a very popular live game where product innovation solely depends on the data. This data-driven game design and development basically depends on the three major approaches, and these approaches generate lots of success for this company through the game in recent times. These approaches are namely, data-driven exploration, data-augmented ideation, and data-informed validation. Each step has a different utility of data in the game development process. These three approaches consider all possible sources of data from the users and use that data as feedback into the data development process. These data sources are playtest data, in-game data, market research data, communication data between the players, comments on the game in the community, game review data, and data from other interactive platforms. All these heterogeneous data about the game help the developer to collect feedback about all the product aspects and help to innovate new modules or new stages of the game. These heterogenous data help the developer to find several design alternatives, or these vast data help the developer to explore several possibilities in the game design. The data-augmented ideation involves the interplay of design creativity of slight modification in the obtained design alternatives for making it more interesting and gamer friendly. The improvement of the design alternatives is done based on the analysis of the data. Once these designs are refined based on the data analysis, finally, the players' interaction data is used for validation of the modified design alternative. This stage is known as the data-informed validation and the most important part of the entire data-driven product design process. This process is an iterative and cyclic process, where with time, more useful interaction data is generated for future development of the product. This case indicates that the proper usage of the data based on a robust strategy can be useful for the product design process.

3.5.2 Improving the ISKCON visiting experience based on text mining

International Society for Krishna Consciousness (ISKON) is known for its charity and good temples all around the world. It is a religious society and has around 600 temples in 60 countries till 2021. It is known for its charity works all around the world and is also known for spreading the message of Bhagwat Gita, holy Hindu scripture. Even in the famous 2005 Stanford commencement speech of Steve Jobs, founder Apple Inc., the mention of ISKCON was heard. Similarly, the ISKCON, Bengaluru temple, is famous in India for its various charitable activities. The most well-known initiative by this organisation is the Akshaya Patra free midday meal programme for the students (Vinodhini et al., 2019). This organisation is always known for its great temple visit experience with a free one-time meal. The temple management's major goal is to provide a world-class spiritual experience within the temple, and all of this is free of charge, which makes it so appealing. It collects donations by distributing various books to its visitors or to the devotees. This world-class experience is generally maintained by the management based on the feedback collected from the temple visitors and different online review data. Online platforms like TripAdvisor, Google Plus, and Facebook are closely monitored by an information technology team, and reviews about the temple are gathered. On the other side, offline feedback is collected through feedback forms and feedback registers available at different points of the temple. The temple information technology team classifies the reviews into four categories: positive, negative, neutral, and mixed. The reviews are analysed weekly to monitor trends in the frequency of positive and negative feedback. Further, they use text mining to find out the pain points of the visitors, and this information is used to improve the overall experience of the temple visit.

3.6 CHALLENGES OF DATA-DRIVEN PRODUCT DESIGN AND DEVELOPMENT

The main challenges for modern product designing are the lack of infrastructure for data management (Cantamessa et al., 2020). This is also a major problem for integrating multiple types of data sources, which may help to consider multiple design aspects of the product design (Briard et al., 2023). Required expertise in data analysis is another major concern for data-driven product design. The correct design analysis is required to find out useful insights about the product design or trend in the consumer feedback and responses. This needs a good data analysis team.

3.7 SUMMARY

In modern times, product design and development are a multidimensional problem. The user or the consumers should be at the centre of the product design and development process as they are the judge of a product. The development of a product should consider both quantitative data and qualitative data from the consumers. The quantitative data tells the story of increasing or declining numbers of sales and can give the information about the particular features that are liked by the users or what are the features they are looking for. However, the qualitative data may give answers to why a particular feature is not liked by the user or what the problems that consumers are facing while using the product. Also, the qualitative data may tell the story which may be improved.

BIBLIOGRAPHY

Alston, F., & DeKerchove, E. M. (2023). *Inclusion by Design: Future Thinking Approaches to New Product Development*. CRC Press.

Andreasen, M. M., Hansen, C. T., & Cash, P. (2015). *Conceptual Design*. Springer.

Bean, R. (2023). Has progress on data, analytics, and AI stalled at your company? *Harvard Business Review*.

Branstad, A., & Solem, B. A. (2020). Emerging theories of consumer-driven market innovation, adoption, and diffusion: A selective review of consumer-oriented studies. *Journal of Business Research*, *116*, 561–571.

Briard, T., Jean, C., Aoussat, A., & Véron, P. (2023). Challenges for data-driven design in early physical product design: A scientific and industrial perspective. *Computers in Industry*, 145, 103814.

Brown, T. (2008). Design thinking. *Harvard Business Review*, *86*(6), 84.

Cantamessa, M., Mntagna, F., Altavilla, S., Casagrande-Seretti, A. (2020) Data-driven design: The new challenges of digitization on product design and development. *Design Science*, 6, e27, doi: 10.1017/dsj.2020.25

Codini, A. P., Abbate, T., & Petruzzelli, A. M. (2023). Business model innovation and exaptation: A new way of innovating in SMEs. *Technovation*, *119*, 102548.

Cross, N. (2023). *Design Thinking: Understanding How Designers Think and Work*. Bloomsbury Publishing.

Dunne, D., & Martin, R. (2006). Design thinking and how it will change management education: An interview and discussion. *Academy of Management Learning & Education*, *5*(4), 512–523.

Fenko, A., & van Rompay, T. J. (2018). Consumer-driven product design. In *Methods in Consumer Research, Volume* 2 (pp. 427–462). Woodhead Publishing.

Holmlid, S. (2007). *Interaction Design and Service Design: Expanding a Comparison of Design Disciplines*.

Jreissat, M., Isaev, S., Moreno, M., & Makatsoris, C. (2017). Consumer driven new product development in future re-distributed models of sustainable production and consumption. *Procedia CIRP*, *63*, 698–703.

Nguyen, J., Ferraro, C., & Sands, S. (2020). Similarity over difference: How congruency in customer characteristics drives service experiences. *Journal of Business Research*, *121*, 592–603.
Sinek, S. (2011). *Start With Why: The Inspiring Million-Copy Bestseller That Will Help You Find Your Purpose*. Penguin UK.
Ulrich, K. T., Eppinger, S. D., & Yang, M. C. (2008). *Product Design and Development* (Vol. 4, pp. 1–3). McGraw-Hill higher education.
Veryzer, R. W., & Borja de Mozota, B. (2005). The impact of user-oriented design on new product development: An examination of fundamental relationships. *Journal of Product Innovation Management*, *22*(2), 128–143.
Vinodhini, R., Vigneswaran, S.R., & Dinesh Kumar, U., (2019) *Enhancing Visitor Experience at ISKON Using Text Analytics*. Harvard Business Publishing.
Werder, K., Seidel, S., Recker, J., Berente, N., Gibbs, J., Abboud, N., & Benzeghadi, Y. (2020). Data-driven, data-informed, data-augmented: How Ubisoft's ghost recon wildlands live unit uses data for continuous product innovation. *California Management Review*, *62*(3), 86–102.
Williams, A. (2007). Product service systems in the automobile industry: Contribution to system innovation? *Journal of cleaner Production*, *15*(11-12), 1093–1103.

Chapter 4

AI and ML-based decision models and their effective utilisation across interdisciplinary applications

R. Kanthavel, R. Dhaya and R. Adline Freeda

4.1 INTRODUCTION TO AI AND ML DECISION MODELS

Rapidly developing technologies like machine learning (ML) and artificial intelligence (AI) have revolutionised many facets of technology and decision-making. ML is a subset of AI that focuses on the creation of algorithms that let computers learn from and make predictions or judgements based on data. AI is the broad category that includes a variety of methods and algorithms that allow machines to mimic human intelligence [Çubukçu et al., 2024]. When combined, these technologies offer strong tools for pattern recognition, complex dataset analysis, and informed decision-making in a variety of fields. The origins of AI and ML may be traced back to the middle of the 20th century. The goal of early AI research was to build machines that could do activities like language comprehension and problem-solving that required human intelligence. Simple algorithms capable of performing classification and linear regression tasks were the first to evolve into ML [Chakraborty and Pillai, 2022]. More complex models, such as neural networks and deep learning, have been developed throughout the years as a result of improvements in processing power, data accessibility, and algorithmic methodologies. These developments have greatly increased the capabilities of AI and ML systems [Wang et al., 2017]. A number of fundamental ideas and methods form the basis of AI and ML:

Supervised Learning: This involves applying labelled datasets for model training, where the target variable or output is known. Linear regression, logistic regression, and support vector machines are examples of common algorithms [4].

Unsupervised Learning: Entails using unlabelled data to train models with the intention of finding underlying patterns or structures. Principal component analysis and dimensionality reduction (e.g., k-means) are common techniques in data analysis.

Reinforcement Learning: Involves using rewards and penalties to teach models to make decisions by identifying and punishing desired

 DOI: 10.1201/9781003621102-4

behaviour. This methodology finds frequent application in autonomous systems, robotics, and games [Sarker, 2022a].

Deep Learning: A branch of ML that models complicated patterns in data by employing deep neural networks, which include many layers. It works especially well for picture and speech recognition applications.

Natural language Processing (NLP): Entails creating algorithms that are capable of producing, understanding, and interpreting human language. Chatbots, sentiment analysis, and language translation are a few examples of applications [Kanthavel et al., 2022].

Decision models based on AI and ML have become essential tools in today's technology, fostering efficiency and creativity in a wide range of industries. With these models, businesses can process vast amounts of data, automate difficult operations, and produce highly accurate forecasts. Businesses, healthcare, finance, education, and many other industries stand to gain greatly from the capacity to use AI and ML for decision-making, underscoring the importance of these technologies in today's technological environment [Sarker, 2022b]. In order to explore the transdisciplinary applications of AI and ML decision models, this introduction offers a basic grasp of them. Examining the fundamental ideas, background, and effects of these technologies can help us understand how they are changing our environment and how they have the ability to be revolutionary.

4.2 INTERDISCIPLINARY APPLICATIONS OF AI AND ML-BASED DECISION MODELS

Decision models powered by AI and ML have the ability to revolutionise a wide range of interdisciplinary sectors by bridging gaps between previously isolated domains. The following are a few important transdisciplinary uses:

Figure 4.1 Interdisciplinary applications of AI- and ML-based decision models.

i. Medical Care and Medication:
- **Diagnostic Decision Support:** AI helps with pathology (e.g., tissue sample analysis) and medical imaging analysis (e.g., tumour detection in radiology scans).
- **Personalised medicine:** By using clinical and genetic data to anticipate patient reactions to therapies, ML algorithms enable customised treatment regimens [Adline R. Freeda et al., 2024].
- **Therapeutic Discovery:** By evaluating chemical structures, forecasting therapeutic efficacy, and spotting possible adverse effects, AI speeds up the creation of new drugs.

ii. Environmental Science:
- **Climate Modelling:** AI analyses environmental effects, forecasts climatic trends, and suggests mitigation and adaptation measures.
- **Ecological Monitoring:** ML uses satellite imagery analysis to track changes in habitat, biodiversity, and rates of deforestation.

iii. Personalised Learning:
- **AI improves** learning results by customising the pace and content of lessons to each unique student.
- **Academic performance** is predicted, at-risk pupils are identified, and interventions for student success are suggested via educational analytics using ML [Afzaliseresht et al., 2020].

iv. Smart Cities and Urban Planning:
- **AI traffic management** shortens commute times by coordinating signals, anticipating patterns of traffic congestion, and optimising traffic flow.
- **Energy Efficiency:** ML algorithms maximise building and urban infrastructure energy use, hence fostering sustainability [Leelavathy et al., 2023].
- **Public Safety:** AI uses crime data analysis to identify hotspots and efficiently distribute resources for preventing crimes.

v. Psychology and Social Sciences:
- **Sentiment Analysis:** ML analyses social media data to forecast trends and determine how the public feels about particular subjects [Dhaya and Kanthavel, 2021a].
- AI uses behavioural modelling to simulate human behaviour in order to comprehend social dynamics, customer preferences, and decision-making processes.

vi. Robots and Self-Driving Systems:
- **Industrial automation:** AI-driven robots carry out intricate operations in shipping, manufacturing, and assembly.
- **Healthcare Robotics:** By enhancing accuracy and speeding up recovery periods during minimally invasive treatments, autonomous surgical robots help [Anuradha et al., 2021].

vii. Ethics and the Law:
- **Legal Decision Support:** To help attorneys with legal research and case outcome prediction, AI examines legal texts, case histories, and precedents.
- **AI Ethics:** Multidisciplinary research aims to create AI systems that are just, open, and responsible while taking ethical considerations into account when making decisions.

AI and ML-based decision models use data-driven insights, pattern recognition, and predictive skills in each of these domains to improve decision-making, allocate resources optimally, and spur innovation. These applications' interdisciplinary nature highlights how AI and ML have the ability to completely change how problems are treated and resolved in a variety of fields, which will ultimately result in more effective, efficient, and informed decision-making [Kanthavel et al., 2023].

4.3 CHALLENGES AND CONSIDERATIONS IN IMPLEMENTING AI AND ML-BASED DECISION MODELS

Decision models based on AI and ML must be implemented, which presents a number of organisational, ethical, operational, and technical obstacles. As AI models mostly rely on data for training and inference, first and foremost, making sure the quality and quantity of data is crucial. Prejudicial results can be prevented by using data that is precise, pertinent, and representative of the issue domain. Another crucial problem is choosing the right models and algorithms, striking a compromise between simplicity and interpretability by sacrificing accuracy for complexity [Dhaya, R., R. Kanthavel, 2021b]. Models must process massive amounts of data quickly and effectively, which makes scalability difficult and frequently necessitates the use of strong infrastructure and optimisation techniques.

The recruitment and retention of highly qualified personnel with AI, ML, and domain-specific expertise are among the organisational obstacles. Organisations must overcome resistance to AI adoption and maintain alignment with current procedures and user acceptability, which makes change management essential [Agrawal and Srikant, 1994a]. Due to the dangers that biased data and algorithms pose to accountability and fairness, ethical issues are quite important. Robust procedures are necessary to safeguard sensitive data and adhere to legal frameworks such as General Data Protection Regulation (GDPR) and Health Insurance Portability and Accountability Act (HIPAA) due to privacy and security issues.

Making sure AI models are comprehensible and interpretable is one of the operational considerations, particularly in regulated industries where openness is essential. In order to keep the model accurate and relevant over time, maintenance entails tracking model performance, spotting drift, and guaranteeing updates. Gaining a competitive edge in today's data-driven environment, utilising AI and ML for decision-making, and driving innovation all depend on proactive management of these issues and problems [Alazab et al., 2018a]. For organisations to successfully use AI- and ML-based decision models, they must create complete plans, invest in the right personnel and infrastructure, and maintain ethical standards.

4.4 ENSURING ETHICAL AND RESPONSIBLE USE OF AI AND ML

As AI and ML are incorporated into more facets of society, it is critical to ensure their ethical and responsible use. Several important tenets and procedures are involved in addressing ethical considerations:

Assimilated Into a Variety of Social Contexts: Taking ethical issues into account requires a number of important tenets and procedures:

- Equity and Mitigation of Bias: Putting methods in place to identify and reduce biases in datasets and algorithms to guarantee equitable treatment for a range of demographic groups.
- Systematically checking the fairness of AI systems and assessing the results to find and fix any possible biases [Kanthavel and Dhaya, 2022].

Explain Ability and Transparency:

- Making algorithms and their results comprehensible to stakeholders would ensure transparency in AI decision-making processes.
  - Explaining AI-driven choices, particularly in crucial or high-stakes fields like banking or healthcare.

Data Security and Privacy:

- Protecting people's right to privacy and making sure that data protection laws (such as the California Resident Privacy Rights (CCPA) and GDPR) are followed [Abonamah and Abdelhamid, 2024].
- Putting strong access controls, anonymisation methods, and data encryption into place to safeguard private data that AI systems use.

Governance and Accountability:

- Establishing precise governance and accountability frameworks for the application of AI and ML.
- Assigning roles and duties for the supervision, observation, and auditing of AI systems to guarantee adherence to legal and ethical requirements.

Design with Humans in Mind and User Consent:

- Putting human values front and centre while designing AI systems to make sure they meet social norms and expectations.
- Acquiring users' informed consent for the gathering, processing, and usage of their data in AI applications.

Evaluation and Mitigation of Risks

- Identifying any ethical concerns and ramifications of AI deployments by carrying out comprehensive risk assessments.
- Creating plans to reduce hazards and guarantee that AI systems function morally and securely in practical settings.

Ongoing Awareness and Education:

- Delivering continuing instruction on ethical issues and best practices to data scientists, stakeholders, and AI developers.
- Educating the public at large about the moral ramifications of AI and ML technology.

Cooperation and a Multidisciplinary Strategy:

- Fostering cooperation to create ethical frameworks and guidelines amongst ethicists, technologists, legislators, and stakeholders [Mittermaier et al., 2023].
- Integrating a range of viewpoints and knowledge into decision-making procedures in order to fully handle ethical dilemmas.

Organisations may build trust, reduce risks, and guarantee that these revolutionary technologies are utilised responsibly to maximise benefits to society and minimise potential downsides by incorporating these concepts into AI and ML development and deployment processes [Radhakrishnan et al., 2022]. As AI systems are designed, developed, deployed, monitored, and used in other phases of their lives, ethical issues should be given top priority in order to foster a future in which AI improves human welfare while adhering to moral norms.

4.5 INTEGRATING AI AND ML-BASED DECISION MODELS WITH EXISTING SYSTEMS

The process of smoothly integrating advanced analytics and machine learning capabilities into an organisation's current IT infrastructure and operational workflows is known as integration of AI and ML-based decision models with existing systems [Huerta et al., 2023]. With the help of this integration, decision-making procedures will be improved, jobs will be automated, and data will yield useful insights. The following explains the main elements of this integration:

4.5.1 Technical integration

Data Integrity and Compatibility:

- **Data Assessment:** Start by determining whether the current data sources are compatible with the needs of AI and ML. This entails assessing the format, quality, and accessibility of data.

- **Data Preparation:** Make sure your data satisfies the requirements needed for AI model training and deployment by cleaning, normalising, and transforming it [Sangeetha and Dhaya, 2022].
- **Data Integration:** Construct interfaces or pipelines that enable efficient data transfer across current applications, databases, and AI models. Depending on the use case, this could entail batch processing or real-time data streaming.

Model Development and Selection:

- **Model Alignment:** Choose AI and ML models that align with the particular use cases and business goals. Consider variables such as the problem's complexity, the dataset's size, and the desired degree of interpretability.
- **Development:** Create and train AI models using appropriate frameworks (like TensorFlow and PyTorch) and algorithms (such as neural networks, decision trees, and regression) that perform well with the infrastructure and technology already in place [Agrawal and Srikant, 1994b].

Architecture of Deployment:

- **Scalable Infrastructure:** Create an infrastructure that satisfies the performance and scalability demands of AI models. Using cloud computing resources or streamlining on-premises infrastructure for parallel processing and effective resource use are two possible approaches to this [Aha et al., 1991].
- Use microservices or Application Programming Interfaces (APIs) to implement AI features so they may be quickly and simply integrated with current systems. Modular development, deployment, and scalability are made possible by this method.

4.5.2 Integration of organisations

Change Management

Involve stakeholders as early in the process as possible to win their support and resolve any issues they may have with the adoption and integration of AI [Alazab et al., 2018a].

Education and Training: Offer training courses to give staff members the know-how to comprehend and use AI-driven findings.

Cooperation between Teams

Cross-Functional Teams: Encourage cooperation between business units, IT departments, and data scientists to guarantee that AI efforts are in line with long-term corporate objectives [Alazab et al., 2018b].

Constant Feedback: Create feedback loops to collect feedback from stakeholders and end users, enabling iterative changes in integration procedures and AI models.

Security and Privacy of Data:

Observance: Ensure adherence to data protection laws (such as HIPAA and GDPR) by putting strong encryption, access control, and data security mechanisms in place [Zoph et al., 2016].

AI Ethics in Practice: In AI-driven decision-making processes, take ethical factors like bias reduction, fairness, and transparency into account.

4.5.3 Operational management

- Monitoring and Maintenance
  Performance Monitoring: Use dashboards and monitoring tools to keep an eye on the real-time performance of your AI models. To spot irregularities and guarantee peak performance, keep an eye on metrics like accuracy, latency, and throughput.
  Model Maintenance: To keep the model accurate and relevant over time, set up processes for version control, updates, and retraining.

Persistent Enhancement [Ale et al., 2019]

- Iterative Development:
  Adopt an attitude of constant improvement by refining AI models in response to user input, fresh data, and changing business needs.
  Innovation: To generate new product offers, improve consumer experiences, and obtain a competitive edge, investigate potential for innovation by utilising AI capabilities [G.I. Parisi et al., 2019].

In conclusion, incorporating AI and ML-based decision models into current systems necessitates a multifaceted strategy that takes organisational preparedness, technological proficiency, ethical issues, and strategic alignment with corporate objectives into account [Allahyari et al., 2017]. Organisations may leverage AI's revolutionary potential to enhance decision-making, streamline operations, and spur creativity in a quickly changing digital context by tackling these areas holistically [Pattabiraman et al., 2016].

4.6 CASE STUDIES: SUCCESSFUL IMPLEMENTATION OF AI AND ML-BASED DECISION MODELS

4.6.1 Case study 1 Early detection of eye diseases using Google DeepMmind and healthcare

- **Industry:** Healthcare Use Case: Google DeepMind created an AI system to evaluate retinal images and find early indicators of eye conditions such as age-related macular degeneration and diabetic retinopathy.
- **Impact:** By accurately diagnosing eye diseases, the AI model allowed for early intervention and maybe prevented patients' eyesight loss.

4.6.2 Case study 2: Recommendation systems and Netflix:

- **Sector:** Entertainment and Media
- **Application:** Netflix's recommendation engine, which makes individualised content recommendations based on user viewing preferences, past viewing behaviour, and other data, is powered by AI and ML algorithms.
- **Impact:** By providing personalised content recommendations, Netflix was able to increase user engagement and retention, which was a major factor in the company's success in keeping members.

4.6.3 Case study 3: Autonomous Driving with Tesla:

- Tesla uses AI and ML to create autonomous driving capabilities in its cars. These technologies interpret and respond to traffic conditions in real time using deep learning algorithms and computer vision.
- Impact: Established Tesla as a market leader for electric vehicles; it improved vehicle safety, decreased accidents, and cleared the path for upcoming developments in self-driving technology.

4.6.4 Case study 4: Amazon with supply chain management with predictive analytics

- **Sector:** Retail and Online Shopping
- **Application:** Predictive analytics is a feature of Amazon's supply chain management that makes use of AI and ML. To enhance logistics and consumer satisfaction, algorithms forecast demand, optimise inventory levels, and predict product trends.
- **Impact:** Improved delivery times, lower expenses, fewer stock outs, and more operational efficiency all help Amazon meet consumer expectations and keep its competitive edge.

4.6.5 Case study 5: IBM Watson and diagnostics for healthcare

- **Industry:** Healthcare Application: IBM Watson Health uses AI to evaluate patient data, diagnostic imaging, and medical literature to help medical practitioners diagnose and treat patients correctly.
- **Effect:** Using AI-powered insights and suggestions can lead to quicker decision-making in healthcare settings, better patient outcomes, and increased capacity for medical research.

Table 4.1 Case studies

Case study	*Case study*	*Case study*	*Case study*
Google DeepMind and Healthcare	Healthcare	Early detection of eye diseases using AI in retinal scans	High accuracy in diagnosing eye conditions, early intervention, and potential vision loss prevention
Netflix and recommendation systems	Media and entertainment	Personalised content recommendations based on AI and ML	Improved user engagement, increased retention rates, and tailored user experience
Tesla and autonomous driving	Automotive	Development of autonomous driving using AI and ML	Enhanced vehicle safety, reduced accidents, and advancement in self-driving technology
Amazon and predictive analytics	Retail and E-commerce	Predictive analytics in supply chain management	Increased operational efficiency, reduced costs, improved inventory management, and faster delivery
IBM Watson and healthcare diagnostics	Healthcare	AI-driven analysis of medical data for diagnostics	Accelerated decision-making, improved patient outcomes, and enhanced medical research capabilities

4.7 STRATEGIES FOR EFFECTIVE ADOPTION AND DEPLOYMENT OF AI AND ML-BASED DECISION MODELS

Organisations in a variety of industries are increasingly using AI and ML in the current digital era to extract useful insights from huge datasets. An organised strategy that incorporates organisational alignment, ethical considerations, and technological readiness is necessary for the successful adoption and deployment of these technologies.

4.7.1 Significant techniques:

a. Unambiguous Business Goals and Use Case Determination
    - Identify certain business issues or opportunities (e.g., process optimisation, customer experience enhancement, or operational efficiency enhancement) where AI and ML might bring substantial value.
    - Set quantifiable targets to monitor how the use of AI affects company results and make sure that they are in line with strategic goals.

b. Infrastructure Readiness and Data Preparedness
    - Make comprehensive evaluations of current data sources to guarantee their quality, applicability, and quantity for training and validating AI models.
    - Establish strong data governance procedures to ensure data security, integrity, and adherence to legal requirements.
    - Choose platforms and solutions for AI and ML that are compatible with the organisation's scaling needs and technological setup.

c. Collaboration across Functions and Involvement of Stakeholders
    - To ensure that AI activities are in line with corporate aims and goals, encourage cooperation between IT, data science, and business unit teams.
    - Involve important parties early on in the process to gain support, resolve issues, and advance a culture of data-driven decision-making.
    - Create interdisciplinary teams that efficiently handle complex challenges and spur innovation by combining technical proficiency with domain experience.

d. The Management of Change and Enhancement of Skills
    - Effective change management techniques can help the organisation overcome cultural barriers and encourage a data-driven approach.
    - Offer thorough training courses to upskill staff members in data analytics, AI, and ML, enabling them to make good use of emerging technology.

e. Iterative development, performance monitoring, and pilot testing
   - Start small pilot programs to verify AI models, evaluate viability, and show measurable results before expanding deployments throughout the company.
   - When developing AI, take an iterative approach, regularly improving models in response to user input, fresh data insights, and changing business needs.
   - To ensure ongoing optimisation and alignment with corporate objectives, analyse the efficacy and influence of AI models on business processes by setting up metrics and key performance indicators (KPIs).

f. Governance Structures and Ethical Issues
   - Discuss the ethical ramifications of using AI and ML, such as reducing bias, ensuring justice, and being transparent in the decision-making process.
   - Create governance structures to supervise AI initiatives, guarantee adherence to legal requirements, and control security and privacy threats.

An all-encompassing strategy that incorporates organisational preparedness, technological know-how, and ethical considerations is necessary for the successful adoption and implementation of AI- and ML-based decision models. Organisations may use these tactics to take advantage of AI's transformative potential to improve operations, spur innovation, and gain a competitive edge in the data-driven economy of today.

4.8 CONCLUSION: THE FUTURE OF AI AND ML-BASED DECISION MODELS

Future AI and ML-based decision models have the potential to completely transform corporate operations in the digital era and upend entire industries. More complex and precise predictions will be possible, thanks to algorithmic advancements, especially in deep learning and natural language processing. This will improve decision-making across a wide range of applications. With the ability to provide individualised solutions and maximise operational efficiency, these technologies will continue to spur innovation in the industrial, healthcare, and financial sectors, among other industries. AI and Internet of Things (IoT) integration will result in smarter systems with predictive insights and real-time adaptability. Globally, ethical concerns about data privacy, transparency, and bias reduction will continue to be crucial in determining governance standards and legal frameworks. To successfully negotiate these complications, organisations need to make investments in workforce development and cultivate a culture of responsible AI adoption. The landscape of the future is one in which AI and ML-based decision models respect ethical standards, accelerate growth, and enable enterprises to attain previously unheard-of levels of insight and efficiency.

REFERENCES

Abonamah, A.A., N. Abdelhamid Managerial insights for AI/ML implementation: A playbook for successful organizational integration. *Discover Artificial Intelligence* 4, 22 (2024). doi: 10.1007/s44163-023-00100-5

Afzaliseresht, N, Y. Miao, S. Michalska, Q. Liu, H. Wang From logs to stories: human-centred data mining for cyber threat intelligence. *IEEE Access* 8, 19089–19099 (2020).

Agrawal, R., R. Srikant Fast algorithms for mining association rules. In: *Proceedings of the international joint conference on very large data bases*, Santiago Chile, vol 1215, pp. 487–499 (1994a).

Agrawal, R., R. Srikant Fast algorithms for mining association rules. In: *Proceedings of the international joint conference on very large data bases*, Santiago Chile, vol 1215, pp. 487–499 (1994b).

Aha, D.W., D. Kibler, M.K. Albert. Instance-based learning algorithms. *Machine Learning*, 6(1), 37–66 (1991). doi: 10.1007/BF00153759

Alazab, A., S. Bevinakoppa, A. Khraisat. Maximising competitive advantage on e-business websites: a data mining approach. In: *2018 IEEE conference on big data and analytics (ICBDA), IEEE*, pp. 111–116, (2018b).

Alazab, A., S. Bevinakoppa, A. Khraisat Maximising competitive advantage on e-business websites: a data mining approach. In: *2018 IEEE conference on big data and analytics (ICBDA), IEEE*, pp. 111–116, (2018a).

Ale, L., A. Sheta, L. Li, Y. Wang, N. Zhang Deep learning based plant disease detection for smart agriculture. In: *2019 IEEE globecom workshops (GC Wkshps), IEEE*, pp. 1–6 (2019).

Allahyari, M., S. Pouriyeh, M. Assefi, S. Safaei, E.D. Trippe, J.B. Gutierrez, K. Kochut A brief survey of text mining: classification, clustering and extraction techniques. ArXiv: 1707.02919 (arXiv preprint) (2017).

Anuradha, J., et al. Big data based stock trend prediction using deep CNN with reinforcement-LSTM model. *International Journal of Systems Assurance Engineering and Management*, 2, 1–11 (2021).

Chakraborty, S., S. Pillai AI/ML enabled decision making in facilitating robotic surgery, *2022 International Conference on Decision Aid Sciences and Applications (DASA)*, Chiangrai, Thailand, pp. 42-46, 2022. doi: 10.1109/DASA54658.2022.9765103

Çubukçu, Hikmet Can, Deniz İlhan Topcu, Sedef Yenice Machine learning-based clinical decision support using laboratory data. *Clinical Chemistry and Laboratory Medicine (CCLM)*, 62, 5,793-823(2024). doi: 10.1515/cclm-2023-1037

Dhaya, R., R. Kanthavel. Cloud—based multiple importance sampling algorithm with AI based CNN classifier for secure infrastructure. *Automated Software Engineering*, 28.2, 16 (2021a).

Dhaya, R., R. Kanthavel. Cloud—based multiple importance sampling algorithm with AI based CNN classifier for secure infrastructure. *Automated Software Engineering* 28(2), 16 (2021b).

Freeda, Adline R., et al. *Role of ChatGPT in smart cities, applications, challenges, and the future of ChatGPT*, IGI GLOBAL (2024). doi: 10.4018/979-8-3693-6824-4.ch010

Huerta, E.A., B. Blaiszik, L.C. Brinson et al. FAIR for AI: An interdisciplinary and international community building perspective. *Scientific Data* 10, 487 (2023). doi: 10.1038/s41597-023-02298-6

Kanthavel, R., R. Dhaya. Prediction model using reinforcement deep learning technique for osteoarthritis disease diagnosis. *Computer Systems Science and Engineering* 42.1 (2022).

Kanthavel, R., R. Dhaya, A. Ahilan. AI-based efficient WUGS network channel modeling and clustered cooperative communication. *ACM Transactions on Sensor Networks*, 18, 3 (2022).

Kanthavel, R., M. Margala, S. Siva Shankar, P. Chakrabarti, R. Dhaya, T. Chakrabarti Osteoarthritis detection using deep learning-based semantic GWO threshold segmentation. In: Joby, P.P., Alencar, M.S., Falkowski-Gilski, P. (eds) *IoT based control networks and intelligent systems*. ICICNIS (2023). Lecture Notes in Networks and Systems, vol 789.Springer, Singapore. doi: 10.1007/978-981-99-6586-1_41

Leelavathy, S., et al. Effective traffic model for intelligent traffic monitoring enabled deep RNN algorithm for autonomous vehicles surveillance systems. *2023 4th International Conference on Smart Electronics and Communication (ICOSEC). IEEE*, (2023).

Mittermaier, M., M.M. Raza, J.C. Kvedar Bias in AI-based models for medical applications: challenges and mitigation strategies. *NPJ Digital Medicine*, 6, 113 (2023). doi: 10.1038/s41746-023-00858-z

Parisi, G.I., R. Kemker, J.L. Part, et al. Continual lifelong learning with neural networks: *A review*, *Neural Networks*, 113, 54–71 (2019).

Pattabiraman, P., R. Dhaya, B. Sasidhar, R. Kanthavel IoV: The future of automobiles. In Advances in Computing and Data Sciences: *First International Conference, ICACDS 2016*, Ghaziabad, India, November 11–12, 2016, Revised Selected Papers 1 (pp. 158–164), (2017).

Radhakrishnan, K., D. Ramakrishnan, O. I. Khalaf, M. Uddin, C. L. Chen, C. M. Wu A novel deep learning-based cooperative communication channel model for wireless underground sensor networks. *Sensors*, 22(12), 4475 (2022).

Sangeetha, S. K. B., R. Dhaya Deep learning era for future 6G wireless communications—Theory, applications, and challenges. *Artificial intelligent techniques for wireless communication and networking*, pp. 105–119 (2022).

Sarker, I.H. AI-based modeling: Techniques, applications and research issues towards automation, intelligent and smart systems. *SN Computer Science*, 23(2), 158. (2022a). doi: 10.1007/s42979-022-01043-x

Sarker, I.H. AI-based modeling: Techniques, applications and research issues towards automation, intelligent and smart systems. *SN Computer Science*, 3, 158 (2022b). doi: 10.1007/s42979-022-01043-x

Wang, J., J. Tang, Z. Xu, Y. Wang, G. Xue, X. Zhang, & D. Yang Spatiotemporal modeling and prediction in cellular networks: A big data enabled deep learning approach. In *IEEE INFOCOM 2017-IEEE conference on computer communications*, pp. 1–9. IEEE. (2017).

Zoph, B., V. Quoc Le. Neural architecture search with reinforcement learning. arXiv preprint arXiv:1611.01578 (2016).

Chapter 5

Data-driven analysis and prediction of weld bead geometry in A-TIG welding

Anurag Sharma, Naman Joshi, Malesh Raghuvanshi, Naman Jain, Pratishtha Sharma and Akarsh Verma

5.1 INTRODUCTION

The stainless-steel fabrication industry makes extensive use of tungsten inert gas welding or TIG welding. When a high degree of weld quality or careful consideration of precision in welding operation is necessary, it has emerged as the most frequently acknowledged option for welding procedure. Due to its poor deposition rate, TIG welding's main drawback is its low productivity. An alternative TIG welding procedure called activated-TIG welding was developed in order to get around this restriction. Before welding, a thin coating of activated flux paste was applied to the workpiece surface, improving the penetration depth and uniformity of TIG welds. By reducing the number of weld passes necessary to form the joints, the use of activated fluxes increases productivity by doing away with the necessity for edge preparation in welding plates, which requires two or three passes in traditional TIG welding. The goal of this research project is to anticipate and analyse the weld bead shape in A-TIG welding by using a data-driven method. We want to construct precise prediction models by gathering real-time data from sensors mounted on welding equipment and comparing them with observed weld bead properties [Dagur et al. 2023; Deep et al. 2021 and Donega et al. 2016]. Engineers and welders will be able to consistently and effectively obtain specified weld bead shapes by using these models to optimize welding parameters. This chapter aims to provide an overview of the A-TIG welding process, a review of the relevant literature on data-driven approaches in welding and weld bead geometry prediction, an explanation of our data collection and analysis methodology, a presentation of our findings, and a discussion of the research implications for the welding industry.

5.2 MATERIAL AND METHODS

5.2.1 Base material

- 304 Stainless Steel: Smooth, homogeneous weld beads with a well-defined bead geometry are the usual outcome of welding 304 stainless steel. It is highly weldable and has a low distortion susceptibility,

 DOI: 10.1201/9781003621102-5

which means that it frequently creates consistent bead shapes with few faults. Bead geometry is generally consistent, exhibiting good penetration and fusion, producing robust and long-lasting welds appropriate for a range of applications.
- Austenitic Stainless: Austenitic stainless-steel grades 304, 304H, 304LN, and 316LN are commonly welded to produce a consistent and seamless weld bead shape [Vasudevan et al. 2007]. These materials have good fluidity and weldability, which produces well-defined weld beads with little to no spatter or flaws. Because of the bead geometry's strong penetration, fusion, and reinforcing, it is appropriate for demanding applications where the quality of the weld is crucial.
- 304H Austenitic Stainless Steel: The weld bead shape produced by welding 304H stainless steel is comparable to that of normal austenitic stainless steels, but it has better high-temperature strength. The bead geometry exhibits good penetration and fusion qualities while maintaining uniformity and smoothness. Although small variations in heat input and cooling rates may have an impact on the bead profile because of its high carbon content, overall weld quality is still quite good.

The distinct composition and metallurgical characteristics of each base material impact the final weld bead geometry, influencing elements, including surface polish, reinforcing profile, and bead shape. In order to maximize welding parameters and guarantee the creation of high-quality welds in A-TIG welding applications, it is vital to comprehend these features [Ghetiya et al. 2014; Jurica et al. 2018; Kumar 2019a and Kumar 2019b].

5.2.2 Welding process

To attain the intended weld bead geometry and mechanical qualities, A-TIG welding requires the selection and optimization of numerous welding parameters. Important variables consist of:

- **Welding Current:** The welding arc's current supply affects both the amount of heat input and the depth of penetration into the base material [Magudeeswaran et al. 2014; Singh et al. 2023 and Vidyarthy et al. 2018].
- **Arc Voltage:** Arc stability, droplet transfer, and the total amount of energy input into the weld pool are all impacted by the voltage across the arc.
- **Travel Speed:** The features of the fusion zone and the width of the bead are determined by the speed at which the welding flame is moved along the weld joint.
- **Shielding Gas Composition:** To protect the weld pool from air contamination, inert gases like argon or helium are commonly used in A-TIG welding.

5.2.3 ANN

The architecture and operation of real-world neural networks in the human brain served as the paradigm for artificial neural networks (ANNs), which are computer models. ANNs are frequently utilized in engineering, finance, healthcare, and manufacturing, among other disciplines, for predictive modelling applications. ANNs are a potent tool for capturing intricate correlations between input parameters and output responses in the context of predicting weld bead geometry in A-TIG welding.

- **Input Layer:** The parameters, including welding current, arc voltage, travel speed, and shielding gas composition, are received by the input layer [Moghaddam and Kolahan 2021].
- **Hidden Layer:** Through a network of complex weighted links along with activation processes, neural pathways in layers that remain hidden process the incoming data [Vasudevan et al. 2007; Venkatesan 2014 and Pavan 2020]. The network may discover detailed links along with patterns in the data revaluated to these layers.
- **Output Layer:** Predicted measures of the weld bead geometry, including bead width, penetrating depth, as well as reinforcement height, are generated by the output layer.
- **Model Validation:** Validating the model's performance on untested data is crucial to determining the ANN's generalization capacity after it has been trained on the dataset. To assess the model's predictive accuracy with new information, techniques like cross-validation procedures or dividing the dataset into training and validation sets can be applied.

5.3 RESULTS AND DISCUSSION

While analysing the data, it was observed that the majority of research articles reported the experimental investigation of weld bead geometry using A-TIG welding and showed a graph of all the analyses.

The error, or the discrepancy between the goal values the model predicts, is shown by the x-axis. The error range in this instance is –4.459 to 4.859. The amount of occurrences, or data points, falling into each error bin is shown on the y-axis. The histogram in question has 20 bins. The three lines represent the distribution of the errors for the training, validation, and testing data sets. The error histogram is a useful tool for assessing the performance of a data-driven model. It can help to identify potential problems with the model, such as overfitting or underfitting. In an ideal scenario, the error histogram would be symmetrical and centred around zero, indicating that the model is making predictions that are neither consistently too high nor too low. In the image, the training data has a more symmetrical distribution than the validation and test data sets, which

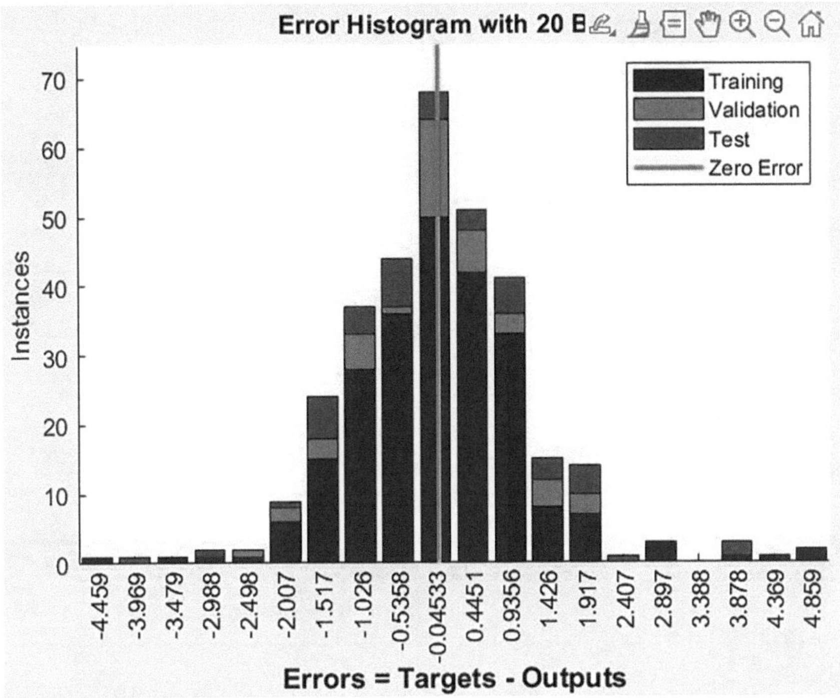

Figure 5.1 Error histogram.

suggests that the model may be overfitting the training data [Ramkumar et al. 2016]. This means that the model is performing well on the training data, but it may not generalize well to unseen data.

A stacked bar histogram showing error distribution across training, validation, and test datasets. The x-axis shows error values ranging from -4.459 to 4.859, while the y-axis shows instances from 0 to 70. The distribution appears roughly normal, centered near zero, with the highest bar reaching about 70 instances. Bars are color-coded in blue (training), green (validation), and red (test), with an orange line indicating zero error.

Here are some other observations that can be made from the image:

The validation and test data sets have a wider distribution of errors than the training data set [Unni and Vasudevan 2020]. This suggests that the model is making more errors on unseen data than on the data it was trained on. The test data set has a slightly higher peak error than the validation data set. This suggests that the model may be generalizing slightly better to the validation data set than to the test data set. Overall, the error histogram in the image suggests that the data-driven model may be overfitting the training data. This is a common problem in machine learning, and it can be mitigated by using techniques such as regularization.

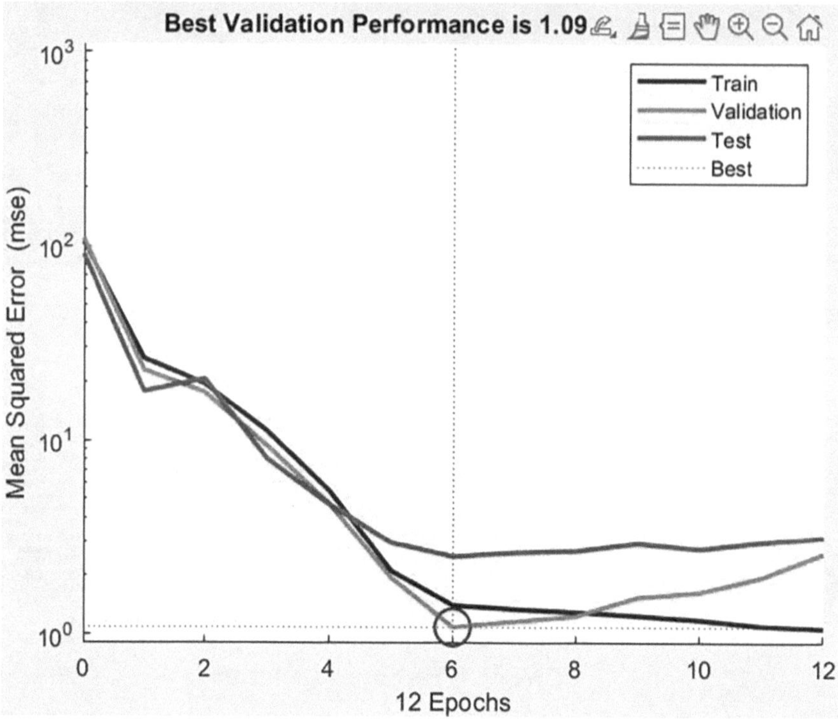

Figure 5.2 Best validation performance.

The x-axis represents the epoch, which is a unit of iteration in machine learning. The y-axis represents the mean squared error (MSE). The text "1.09" in the top right corner indicates the best validation performance achieved by the model. The graph shows how well the model performed on a set of data it was not trained on (validation data) as the training progressed (more epochs). The MSE is a common metric used to measure how different the predictions of the model are from the actual values for the validation data. In this case, a lower MSE indicates a better performance. The graph suggests that the model achieved its best validation performance at around 8 epochs. This means that after 8 epochs of training, the model performed best on the validation data set. It's important to note that the ideal number of epochs can vary depending on the specific data and model used.

Here are some other observations from the image:

> The MSE appears to be decreasing as the number of epochs increases, up to around 8 epochs. This suggests that the model's performance is improving as it is trained on more data. It is difficult to say for certain

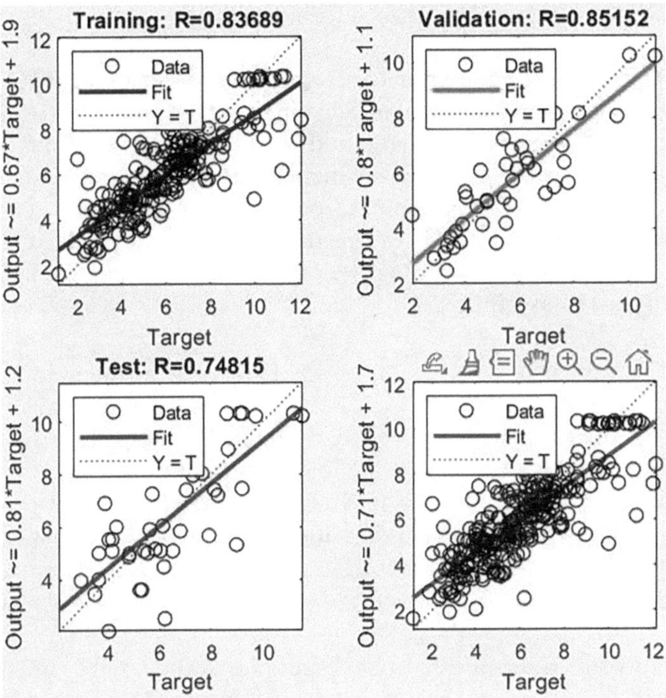

Figure 5.3 Regression plot graph.

from this graph alone whether the model is overfitting the data. Overfitting is a scenario where a model performs well on the training data but poorly on unseen data. To identify overfitting, it would be helpful to see how the model performs on a test data set, which is a completely separate set of data from the training and validation data.

Each data point in the scatter plot represents a weld bead. The x-axis represents the target weld bead geometry, which is the actual measured value. The y-axis represents the predicted weld bead geometry, which is the value predicted by the data-driven model. The text in the top right corner of each plot shows the R-squared value, which is a statistical measure of how well the data points fit a regression line. An R-squared value of 1 indicates a perfect fit, while a value of 0 indicates no linear correlation between the variables. In the training data plot, the R-squared value is 0.83689, which indicates a strong positive correlation between the predicted and target weld bead geometry. In the validation data plot, the R-squared value is 0.85152, which is also a strong positive correlation.

Here are some other observations from the image:

> The data points in both plots are scattered around a diagonal line from the bottom left to the top right. This suggests that there is a positive correlation between the predicted and target weld bead geometry, meaning that as the predicted values increase, the target values also tend to increase. The spread of the data points in the validation data plot is slightly wider than the spread in the training data plot. This suggests that the model may be fitting the training data a little too well, which could lead to overfitting.

The text "Gradient = 0.94155, at epoch 10" refers to the gradient of the loss function at epoch 10. The gradient is a measure of how much the loss function is changing with respect to the model's parameters. In machine learning, the goal is to minimize the loss function by adjusting the model's parameters during training. The text suggests that the gradient has a value of 0.94155 at epoch 10, which may indicate that the loss function is still decreasing at that point in training.

Here are some other observations that can be made from the image:

> The loss function appears to be decreasing as the number of epochs increases. This suggests that the model is improving its fit to the training data as it is trained on more data. It is difficult to say for certain from this graph alone whether the model is overfitting the data. Overfitting is a scenario where a model performs well on the training data but poorly on unseen data. To identify overfitting, it would be helpful to see how the model performs on a test data set, which is a completely separate set of data from the training data.

Box and whisker plots are frequently used to show how data is distributed. The middle quartiles of the data are represented by the box in the plot, outliers are plotted as individual points, and the whiskers extend to the lowest and highest data points within 1.5 times the interquartile range. The data in the figure may reflect measurements of the width, depth, or penetration of the weld bead in the context of data-driven modelling for forecasting weld bead geometry in A-TIG welding. Next, the distribution of these measurements for a certain set of welding conditions would be displayed using the box and whisker plot. Here are some of the things that data-driven models can predict about weld bead geometry in A-TIG welding:

> Weld bead width, weld bead depth, weld bead penetration, dilution, and spatter [Vora et al. 2019; and Vora et al. 2021].
>
> Data-driven models can be used to improve the quality of welds by optimizing welding parameters. For example, a model could be used to predict the weld bead geometry that will result in the strongest weld.

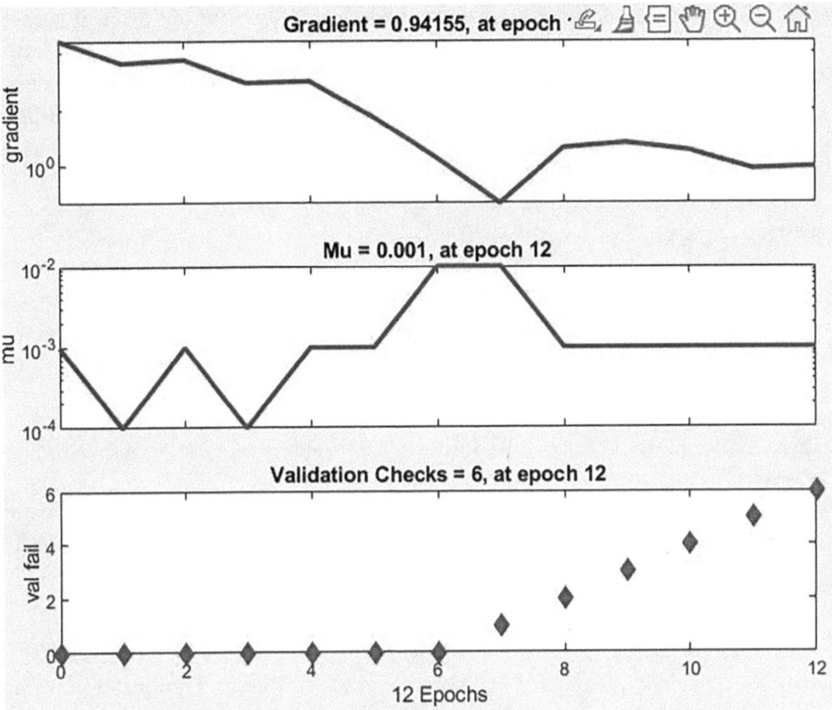

Figure 5.4 Machine learning validation loss graph.

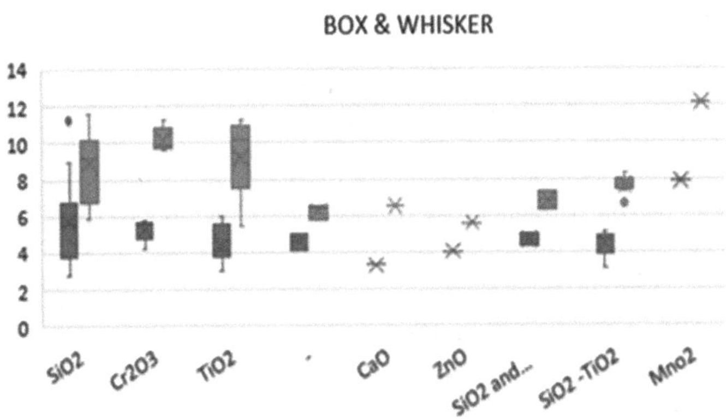

Figure 5.5 Effect of various fluxes.

5.4 CONCLUSIONS

The A-TIG technique can easily and consistently boost the output and quality of traditional TIG welding, which will be useful in the industrial setting. The work's findings demonstrate that using a flux can significantly boost TIG welding's depth of penetration; in this experiment, a roughly 86% increase in penetration depth was noted. According to research on the impact of multicomponent fluxes on A-TIG welding, SiO_2 flux has the greatest influence on increasing penetration depth, yet the impact of a flux containing 100% SiO_2 on penetration depth is not very significant.

REFERENCES

Dagur, H., Kumar, R., Singh, V., Chauhan, S., Deep, A., & Patel, D. (2023). Effect of TIG and activated flux TIG welding processes on weld bead geometry, microstructure, and hardness of SAF 2507 grade super duplex stainless steel joints. *Engineering Research Express*, *5*(3), 035010.

Deep, A., Singh, V., Ashutosh, S., Chandrasekaran, M., & Patel, D. (2021). Performance of weld bead profile during A-TIG welding on nitrogen alloyed stainless steel. *Engineering Research Express*, *3*(4), 045024.

Donegá, T. J., Costa, T. F., Arencibia, R. V., & Vilarinho, L. O. (2016). Comparison of thermal efficiency between A-TIG and conventional TIG welding. *Welding International*, *30*(4), 255–267.

Ghetiya, N., & Pandya, D. (2014, November). Mathematical modeling for the bead width and penetration in activated TIG welding process. In *International Conference on Multidisciplinary Research & Practice* (Vol. 1, No. 7, pp. 247–252).

Jurica, M., Kožuh, Z., Garašić, I., & Bušić, M. (2018, March). Optimization of the A-TIG welding for stainless steels. In *IOP conference series: Materials science and engineering* (Vol. 329, p. 012012). IOP Publishing.

Kumar, H., Ahmad, G. N., & Singh, N. K. (2019b). Activated flux TIG welding of Inconel 718 super alloy in presence of tri-component flux. *Materials and Manufacturing Processes*, *34*(2), 216–223.

Kumar, K., Masanta, M., & Sahoo, S. K. (2019a). Microstructure evolution and metallurgical characteristic of bead-on-plate TIG welding of Ti-6Al-4V alloy. *Journal of Materials Processing Technology*, *265*, 34–43.

Magudeeswaran, G., Nair, S. R., Sundar, L., & Harikannan, N. (2014). Optimization of process parameters of the activated tungsten inert gas welding for aspect ratio of UNS S32205 duplex stainless-steel welds. *Defence Technology*, *10*(3), 251–260.

Moghaddam, M. A., & Kolahan, F. (2021). Activated gas tungsten arc welding process optimization using artificial neural network and heuristic algorithms. https://doi.org/10.21203/rs.3.rs-526655/v1

Pavan, A. R., Arivazhagan, B., & Vasudevan, M. (2020). Process parameter optimization of A-TIG welding on P22 steel. In *Structural Integrity Assessment: Proceedings of ICONS 2018* (pp. 99–113). Springer Singapore.

Ramkumar, K. D., Ramanand, R., Ameer, A., Simon, K. A., & Arivazhagan, N. (2016). Effect of post weld heat treatment on the microstructure and tensile properties of activated flux TIG welds of Inconel X750. *Materials Science and Engineering: A*, *658*, 326–338.

Saha, S., & Das, S. (2020). Effect of polarity and oxide fluxes on weld-bead geometry in activated tungsten inert gas (A-TIG) welding. *Journal of Welding and Joining*, *38*(4), 380–388.

Singh, A., Singh, V., Singh, A. P., Ashutosh, S., & Patel, D. (2023). Welding investigations on mechanical property and microstructure of TIG and A-TIG Weld of Hastelloy C-276. *Engineering Research Express*, *5*(2), 025004.

Unni, A. K., & Vasudevan, M. (2020). Numerical modelling of fluid flow and weld penetration in activated TIG welding. *Materials Today: Proceedings*, *27*, 2768–2773.

Vasudevan, M. (2017). Effect of A-TIG welding process on the welding attributes of type 304LN and 316LN stainless steels. *Journal of Materials Engineering and Performance*, *26*, 1325–1336.

Vasudevan, M., Bhaduri, A. K., Raj, B., & Rao, K. P. (2007). Genetic-algorithm-based computational models for optimizing the process parameters of A-TIG welding to achieve target bead geometry in type 304 L (N) and 316 L (N) stainless steels. *Materials and Manufacturing Processes*, *22*(5), 641–649.

Venkatesan, G., George, J., Sowmyasri, M., & Muthupandi, V. (2014). Effect of ternary fluxes on depth of penetration in A-TIG welding of AISI 409 ferritic stainless steel. *Procedia Materials Science*, *5*, 2402–2410.

Vidyarthy, R. S., & Dwivedi, D. K. (2018). Microstructural and mechanical properties assessment of the P91 A-TIG weld joints. *Journal of Manufacturing Processes*, *31*, 523–535.

Vora, J., Patel, V. K., Srinivasan, S., Chaudhari, R., Pimenov, D. Y., Giasin, K., & Sharma, S. (2021). Optimization of activated tungsten inert gas welding process parameters using heat transfer search algorithm: With experimental validation using case studies. *Metals*, *11*(6), 981.

Vora, J. J., Abhishek, K., & Srinivasan, S. (2019). Attaining optimized A-TIG welding parameters for carbon steels by advanced parameter-less optimization techniques: With experimental validation. *Journal of the Brazilian Society of Mechanical Sciences and Engineering*, *41*, 1–19.

Chapter 6

DFMA integration for leaner and faster LED bulb manufacturing

Prashant Shukla

6.1 INTRODUCTION

In today's dynamic and competitive market landscape, the imperative to trim product development timelines remains paramount for sustaining competitiveness. Design for manufacture and assembly (DFMA) emerges as a robust methodology that not only slashes expenses [1, 2] but also streamlines the intricate process of product development. DFMA entails a systematic deconstruction of a product into its elemental constituents, coupled with an exhaustive analysis of the requisites essential for its fabrication and assembly. This early-stage scrutiny offers invaluable insights into assembly sequences, material logistics, and potential production intricacies, ultimately culminating in cost reduction [3, 4].

DFMA seamlessly integrates the principles of design for X (DFX), encompassing design for manufacturing (DFM) and design for assembly (DFA). While DFA focuses on leveraging efficient fastening mechanisms and minimizing component count to accelerate assembly processes, DFM prioritizes optimizing products for cost-effective manufacturing.

By embedding DFMA into the inception phase, organizations can proactively identify and rectify design flaws, manufacturing bottlenecks, and costly procedures [5]. Despite the upfront investment of time and resources, this proactive approach translates into significant time, cost, and effort savings in subsequent stages. It is noteworthy that the conceptualization phase typically commands 60% to 80% of the total product cost, yet many organizations allocate a mere 5% of their budget to this pivotal phase. By judiciously allocating resources and expertise to the conceptualization phase, organizations can harness the full potential of cost reduction, enhance performance, and ensure project success.

DFMA stands as a highly esteemed and valuable methodology renowned for its capacity to diminish lead times, enhance product quality, and rationalize production costs [6, 7]. Implementing it early in development helps understand product design, assembly, and manufacturing requirements. Design teams and engineers may make educated architectural, material,

DOI: 10.1201/9781003621102-6

component integration, and production feasibility decisions using this information. Understanding assembly sequences and material flows helps companies optimize operations, eliminate redundancies, and simplify manufacturing, improving usability, reliability, and cost. Moreover, DFMA promotes the incorporation of different design principles such as design for assembly and design for manufacturing to enhance the product development process.

Considerable financial investment is essential during the conceptualization phase, given that it constitutes a considerable proportion of the overall expenditure [8, 9]. By allocating sufficient resources and specialized knowledge at the early stages of product development, it becomes feasible to identify potential design flaws, manufacturing constraints, and costly procedures in a proactive manner. This methodology decreases the number of iterations, mitigates risks, and improves project outcomes. By incorporating DFMA into their product development strategies as a foundational element, organizations can effectively harness its complete capabilities. By means of analysis, optimization, and adherence to industry best practices, DFMA enables organizations to attain production processes that are both efficient and cost-effective. Through the simplification of product design and assembly procedures, DFMA ultimately guarantees on-time delivery, improves product dependability, and substantially reduces expenses. It comprises DFM, which prioritizes economical production, and DFA, which emphasizes streamlined assembly [10].

In order to enhance the assembly process (Figure 6.1) [1, 2], a comprehensive strategy is required. To begin, the process entails the establishment of the assembly procedure, the selection of suitable instruments, and the delineation of steps. Following this, an examination of the assembly procedure is conducted in order to identify bottlenecks, inefficiencies, and potential areas for simplification. The utilization of DFA is of paramount importance as it ensures process optimization, component reduction, and improved precision. The process of evaluating design modifications entails determining their impacts on production, cost, quality, and performance [11, 12]. After conducting a feasibility analysis and evaluating the prototype, design modifications are incorporated [13, 14]. Ongoing evaluation and monitoring are critical components in ensuring long-term optimization.

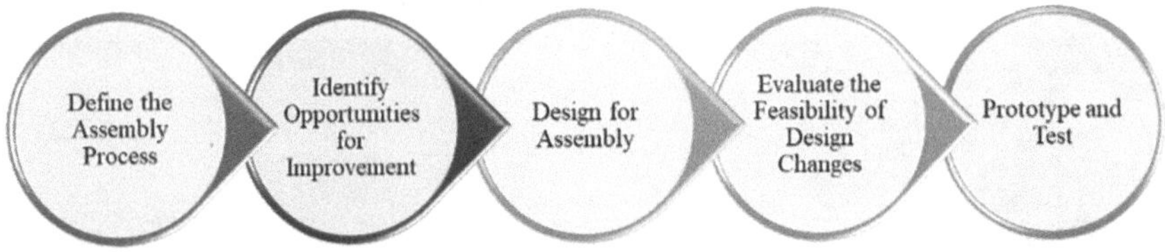

Figure 6.1 Phases of assembly optimization.

6.2 METHODOLOGY

Typically, the DFA procedure involves a systematic analysis of the product's design and assembly techniques. The objective of this evaluation is to differentiate between components that are essential and those that are non-essential (Figure 6.2). The primary objective is to discern the constituents that are essential for the product to operate in an efficient manner, while precisely identifying those that can be substituted or eliminated. This assessment consists of multiple phases, which encompass examining the sequence of assembly for the product, identifying possible obstacles in assembly, and evaluating the simplicity of assembly for the product.

Once the critical and non-essential components have been discerned, the emphasis shifts toward streamlining both the product's design and assembly procedures. The primary objective is twofold: the elimination of non-essential components and the simplification of essential component assembly. This streamlining endeavor culminates in a reduction in the number of components required for product assembly, subsequently leading to diminished assembly duration and cost, alongside the enhancement of overall product reliability.

Hence, the identification of vital and non-essential components constitutes a pivotal phase within the DFA process. It serves the dual purpose of ensuring the product's correct functionality while concurrently simplifying the assembly process and curbing the overall production expenditure.

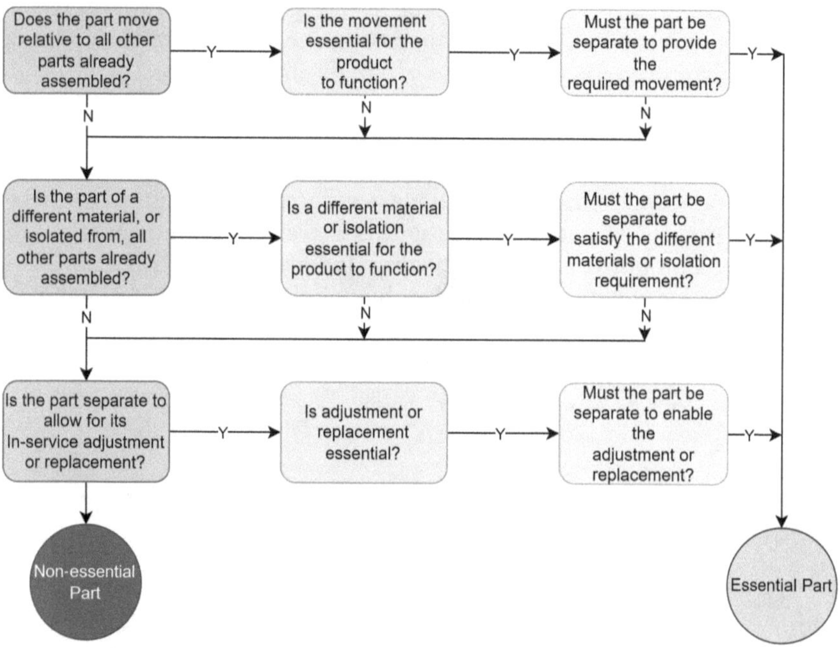

Figure 6.2 Identification of essential and non-essential parts.

6.3 ENHANCING THE ASSEMBLY OF LED BULBS

The 5 main parts that make up a standard LED bulb are listed in Table 6.1 and shown in the schematic in Figure 6.3. The following steps are usually included in the assembly process, as shown in Figure 6.4:

- In Phase 1of LED bulb assembly, the lamp body (Part-2) and connector (Part-1) are integrated. These components are manually joined using machine pressing methods to ensure both mechanical stability and electrical continuity.
- Phase 2 involves machine pressing the lamp body (Part-2) and base (Part-3) together. Standard light fittings seamlessly fit into plastic or metal bases. This process securely connects the lamp body to the base and aligns them.
- Phase 3 involves assembling the LED circuit board, which includes the LED chips and resistors, into the bulb's central component (Part 4). For stable and dependable electrical connections, the circuit board is snugly fastened to the base plate using one or 2 screws.
- Phase 4 includes soldering a connecting wire to the LED board, creating a durable electrical connection that links the board's LED circuits to the power source via the connector.

Following the installation of the LED chips into the circuit board, Phase 5 involves hand-pressing the integrated lens and reflector (Part-6) onto the lamp body (Part-2). The dual-purpose lens shields the LED chips and diffuses light, while the reflector is meticulously designed to channel and enhance the LED chip light. This step completes the construction by precisely positioning and attaching the reflector and lens to the lamp body.

Table 6.1 Analysis of existing LED assembly

S.No	*Parts*	*N. of parts (Np)*	*Theoretical part count*	*No.of interfaces (Ni)*	*Assembly time (Sec.)*	*Remark*
1	Part-1	1	0	2	8	Wire solder
2	Part-2	1	0	3	6	Machined press
3	Part-3	1	0	3	5.5	Machined press
4	Part-4	1	1	2	15	Adhesive, screws, and soldering
5	Part-5	1	0	2	3.5	Screwdrive
6	Part-6	1	1	1	3	Hand press
	Total	**6**	**2**	**13**	**41**	

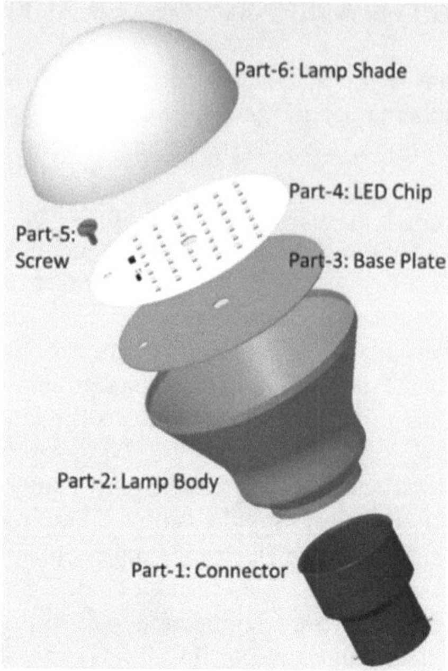

Figure 6.3 Existing LED assembly.

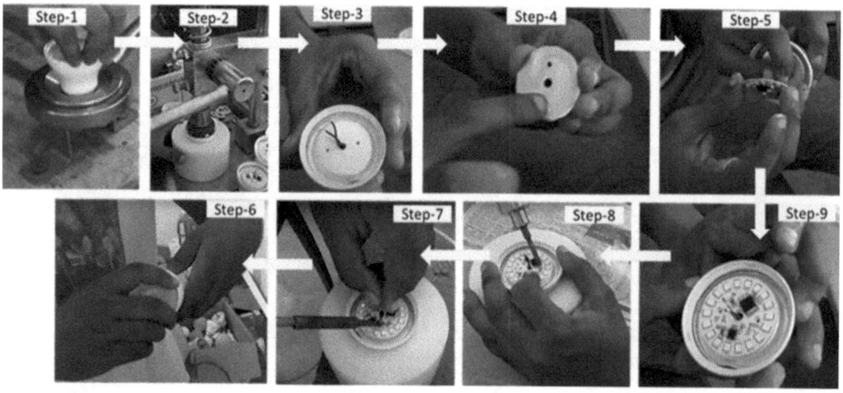

Figure 6.4 LED bulb assembly process.

The comprehensive assembly procedure elucidated previously delineates the precise sequence of actions entailed in the assembly of conventional LED bulbs. Each constituent component is meticulously integrated and firmly fastened to guarantee impeccable electrical continuity, robust mechanical stability, and optimal light emission [15]. By comprehending the intricacies of the assembly process, manufacturers can discern prospects for enhancement. These opportunities may encompass the automation of certain tasks or the introduction of design refinements aimed at simplifying production processes. Such measures contribute to the rationalization of LED bulb manufacturing, ultimately augmenting overall efficiency and elevating the quality standards associated with LED bulb production [16].

The existing design has 5 parts (N_p) and 10 interfaces (N_i), and the assembly process takes 37.5 seconds per piece. The assembly analysis reflects that the total theoretical part count is 2.

$$\text{Theoretical part count efficiency} = \frac{\text{Theroetical min.no.of parts}(\text{Nm})}{\text{Total no.of parts}(N_p)} \times 100 \qquad \text{(Eq.1)}$$

A theoretical component count [17, 18] efficiency of at least 60% [19, 20] is required for any assembly to be considered efficient. The efficiency of the current LED is 33.33 percent.

6.3.1 Reviewing the DFA aspects of LED bulb design

The present design consists of 5 components (N_p) and 10 interfaces (N_i), and the assembly procedure takes 37.5 seconds per unit. The study of the assembly reveals that the theoretical total number of parts is 2.

$$\text{DFA index} = \frac{N_m t_m}{t_a} \times 100 \qquad \text{(Eq.2)}$$

$$\text{And DFA complexity factor} = \sqrt{(\Sigma N_p X \Sigma N_i)} \qquad \text{(Eq.3)}$$

where t_m is the minimum assembly time(sec.)/parts, i.e., 3 seconds, and t_a is the estimated total assembly time (sec.). Hence, the DFA complexity factor is 8.83, and DFA index is 14.63.

6.3.2 DFA analysis of new LED design

The LED lamp's revised design, as illustrated in Figure 6.5, integrates a range of technological advancements with the dual purpose of augmenting its operational capabilities and manufacturing efficacy. A rectangular LED chip, signifying the transition to a surface-mounted LED (SMD) chip, is a notable advance. There are usually several LED elements on the semiconductor's surface, which boosts light output and improves heat dissipation, leading to better performance. The benefits of the redesigned connection, which include switching from a cup to a flat disk form, must also be thoroughly considered. By virtue of its flat disk design, mechanical stability is greatly improved, guaranteeing a solid connection between the power supply and

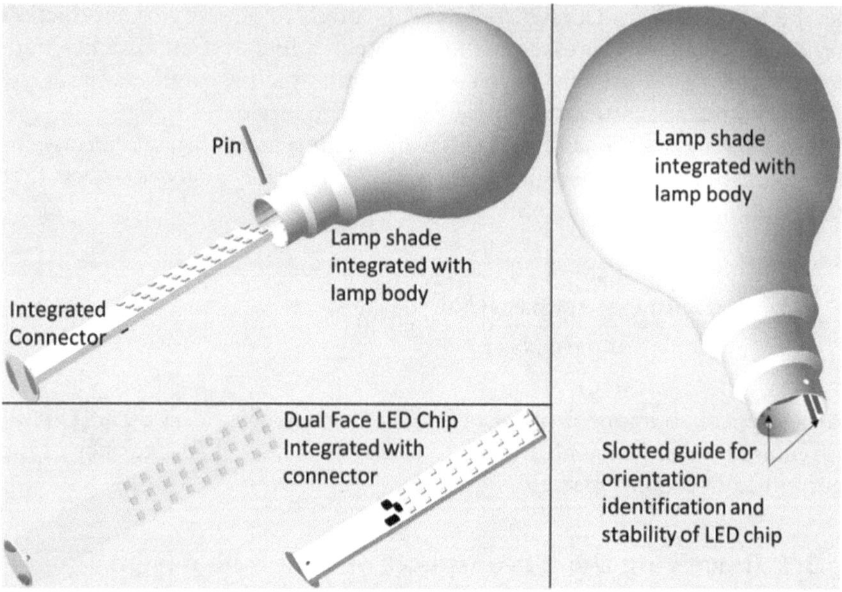

Figure 6.5 New assembly of LED bulb.

the LED chip. Implementing this design modification has the potential to greatly enhance dependability while also decreasing the likelihood of electrical disconnection and operational problems.

6.4 RESULTS AND DISCUSSION

The incorporation of DFMA principles into the LED bulb's design has yielded several noteworthy enhancements, benefiting both its manufacturing and assembly facets. A comprehensive DFA analysis, as illustrated in Table 6.2 and Figure 6.6, offers an intricate juxtaposition of the incumbent design and the new iteration. A prominent enhancement involves the reduction in the overall parts count, dwindling from 6 to a mere 3 components. This reduction serves to streamline the manufacturing process by simplifying the assembly with a decreased number of constituent parts. Additionally, it holds ramifications for inventory management, as well as cost reduction endeavors since a reduced number of components necessitates sourcing and tracking fewer parts.

Theoretically, the redesign has resulted in a part count efficiency of 100%, which means that all of the components are now well-used and contribute to the overall design. A more effective utilization of resources and materials is assured by meticulous optimization, which guarantees the removal of any superfluous or redundant components. Reducing the number of interfaces

Table 6.2 Assembly analysis of new design

Parts	*N. of parts (Np)*	*Theoretical part count*	*No.of interfaces (Ni)*	*Assembly time*
LED chip incorporating a built-in connector	1	1	2	0
Integrated lamp body and lamp shade	1	1	2	5
Lock pin	1	1	2	7
Total	3	3	6	12
DFA Index			75	
DFA Aacomplexity fact			4.24	

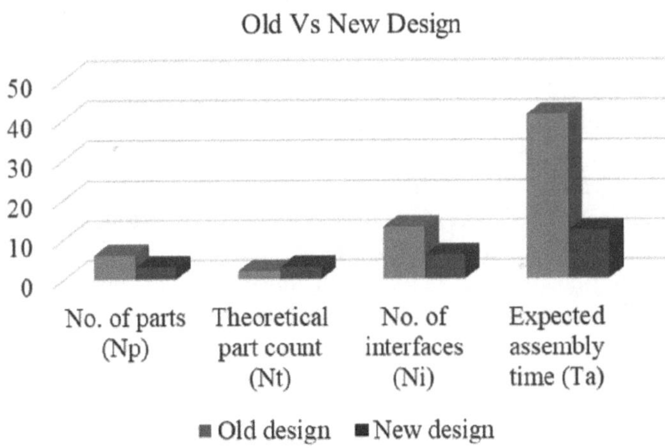

Figure 6.6 Old Vs new design.

from 13 to 6 is an impressive feat. Reducing the amount of interconnections, which are the locations where various components interact, suggests a simpler design with less likelihood of connection difficulties. Reduced performance concerns and increased customer satisfaction are the results of this improvement's impact on the LED bulb's reliability and longevity.

The influence of the DFA exercise is unmistakably apparent (as depicted in Figure 6.7) through the remarkable reduction in manual assembly time, plummeting from 41 seconds to a mere 12 seconds, signifying an impressive 68.3% decrease. This reduction predominantly emanates from the streamlined assembly process, a consequence of the reduced count of components and interfaces [21]. In addition to reducing the probability of assembly errors, the streamlined assembly process also expedites the production of high-quality goods [22].

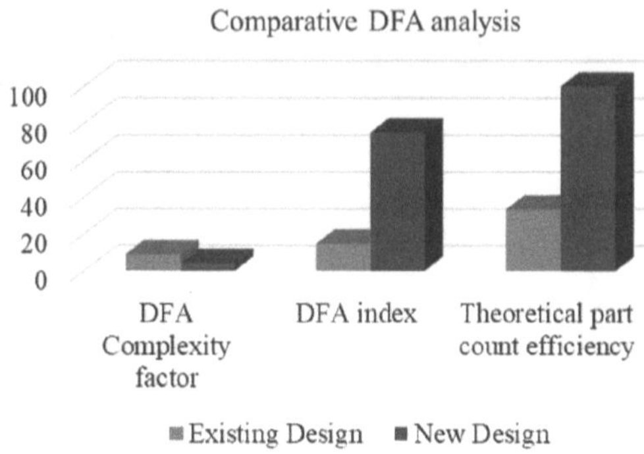

Figure 6.7 Comparative DFA analysis.

The aforementioned results provide additional evidence of the DFA analysis's efficacy. There has been a significant decrease in the overall complexity of the assembly process, as indicated by the DFA complexity factor dropping from 8.83 to 4.24. Moreover, there has been a notable improvement in assembly simplicity and manufacturability, as shown by the increase in the DFA index from 14.63 to 75. In this specific case study, the advantages of integrating DFMA are clearly demonstrated through the overall success achieved. This success encompasses cost reduction, enhanced product quality, improved manufacturability, and optimized design. DFMA assumes a pivotal position in fostering innovation and executing sustainable manufacturing practices through its emphasis on simplification, efficiency, and dependability.

6.5 CONCLUSIONS

There has been a significant decrease in the overall complexity of the assembly process, as indicated by the DFA complexity factor dropping from 8.83 to 4.24. Moreover, there has been a notable improvement in assembly simplicity and manufacturability, as shown by the increase in the DFA index from 14.63 to 75. In this specific case study, the advantages of integrating DFMA are clearly demonstrated through the overall success achieved. This success encompasses cost reduction, enhanced product quality, improved manufacturability, and optimized design. The primary objective of the DFMA methodology is to merge constituent parts into a cohesive entity while removing superfluous elements. By means of this optimization, the assembly process is not only streamlined, but manufacturers also yield cost efficiencies. Companies have the ability to increase profitability through cost savings in production achieved through the elimination of redundant components. The substantial potential for these cost reductions to have a collective effect, especially in the context of mass-producing LED bulbs, can grant organizations a competitive edge.

Companies must foster collaboration between their manufacturing and design teams, which must result in a comprehensive redesign of the entire design process, in order to maximize the benefits of DFMA. Although larger enterprises with more intricate organizational structures may encounter obstacles during this collaborative endeavor, the potential advantages surpass the challenges. In light of concerns surrounding climate change, an increasing number of contemporary businesses are placing a premium on carbon footprint reduction and the implementation of sustainable business practices. Attaining cost reductions and promoting sustainable manufacturing practices can be significantly facilitated by the integration of the DFMA methodology and design for environment (DFE) guidelines. Subsequent research endeavors may encompass cost-benefit analyses and sustainability evaluations of the initial and substantially revised product models, with the aim of quantifying the consequences of conformity with DFE guidelines. Delving deep into the trade-off between cost and environmental benefits in product design could be better understood by thoroughly evaluating sustainability versus cost.

In summary, the implementation of DFMA principles may yield cost savings, enhance manufacturing efficiency, and grant organizations a competitive advantage. By integrating sustainable design principles, organizations can harmonize their activities with ecological goals and fortify their market standing. By combining the DFMA and DFE methodologies, businesses can contribute to a more sustainable future by realizing economic and ecological benefits.

REFERENCES

[1] G. Boothroyd, "Product design for manufacture and assembly," *Comput. Des.*, vol. 26, no. 7, pp. 505–520, Jul. 1994, doi: 10.1016/0010-4485(94)90082-5.

[2] G. Formentini, N. Boix Rodríguez, and C. Favi, "Design for manufacturing and assembly methods in the product development process of mechanical products: a systematic literature review," *Int. J. Adv. Manuf. Technol.*, vol. 120, no. 7–8, pp. 4307–4334, Jun. 2022, doi: 10.1007/S00170-022-08837-6/TABLES/7.

[3] M. Tucci, F. De Carlo, O. Borgia, and N. Fanciullacci, "Accelerated life tests as an integrated methodology for product design, qualification and production control: a case study in household appliances," vol. 2, no. 1, pp. 112–127, Jan. 2014, doi: 10.1080/21693277.2014.893415. http://mc.manuscriptcentral.com/tpmr

[4] P. Suresh, S. Ramabalan, and U. Natarajan, "Integration of DFE and DFMA for the sustainable development of an automotive component," vol. 9, no. 2, pp. 107–118, Mar. 2015, doi: 10.1080/19397038.2015.1096313.

[5] D. R. Mandasari et al., "Design and optimization of brushless DC motor for electric boat thruster," *Evergreen*, vol. 10, no. 3, pp. 1928–1937, 2023, doi: 10.5109/7151773.

[6] G. Boothroyd, P. Dewhurst, and W. A. Knight, "Product design for manufacture and assembly: Third edition," *Prod. Des. Manuf. Assem. Third Ed.*, pp. 1–671, Jan. 2010, doi: 10.1201/9781420089288/PRODUCT-DESIGN-MANUFACTURE-ASSEMBLY-PETER-DEWHURST-WINSTON-KNIGHT-GEOFFREY-BOOTHROYD.

[7] O. Canciglieri, J. P. B. Kovalchuk, M. Rudek, and T. M. De Souza, "Development of White Goods parts in a concurrent engineering environment based on DFM/DFA concepts," *Adv. Concurr. Eng.*, pp. 443–452, 2010, doi: 10.1007/978-0-85729-024-3_47/COVER.

[8] S. K. Das, P. Yedlarajiah, and R. Narendra, "An approach for estimating the end-of-life product disassembly effort and cost," *Int J Prod Res*, vol. 38, no. 3, pp. 657–673, Jan. 2000, doi: 10.1080/002075400189356.

[9] J. Zhang and S. C. Sarin, "Simultaneous determination of design, pricing, and an end-of-life option of a product," *J. Remanufacturing*, vol. 12, no. 3, pp. 357–383, Nov. 2022, doi: 10.1007/S13243-022-00114-9/METRICS.

[10] P. Selvaraj, P. Radhakrishnan, and M. Adithan, "An integrated approach to design for manufacturing and assembly based on reduction of product development time and cost," *Int. J. Adv. Manuf. Technol.*, vol. 42, no. 1–2, pp. 13–29, May 2009, doi: 10.1007/S00170-008-1580-8/METRICS.

[11] S. Schmidt, "Preventive optimisation of costs and quality for the total life cycle – Design for manufacture, assembly, service, environment (DFMA)," *SAE Tech. Pap.*, Nov. 1998, doi: 10.4271/982166.

[12] B. Raucent, C. Nederlandt, and D. A. Johnson, "Plastic snapfit fastener design," *Int. J. Adv. Manuf. Technol.*, vol. 14, no. 3, pp. 185–190, 1998, doi: 10.1007/BF01188414/METRICS.

[13] N. Boorsma, R. Balkenende, C. Bakker, T. Tsui, and D. Peck, "Incorporating design for remanufacturing in the early design stage: a design management perspective," *J. Remanufacturing*, vol. 11, no. 1, pp. 25–48, Apr. 2021, doi: 10.1007/S13243-020-00090-Y/TABLES/4.

[14] L. Lindkvist Haziri and E. Sundin, "Supporting design for remanufacturing – A framework for implementing information feedback from remanufacturing to product design," *J. Remanufacturing*, vol. 10, no. 1, pp. 57–76, Apr. 2020, doi: 10.1007/S13243-019-00074-7/TABLES/5.

[15] S. Agarwal, M. Tyagi, and R. K. Garg, "Circular economy reinforcement to diminish GHG emissions: A grey DEMATEL approach," *Evergreen*, vol. 10, no. 1, pp. 389–403, 2023, doi: 10.5109/6781099.

[16] W. Izdebski, S. Serafin, and V. Tarasevych, "The application of the life cycle assessment method in managing the logistics of municipal waste collection," *Evergreen*, vol. 10, no. 3, pp. 1242–1252, 2023, doi: 10.5109/7148445.

[17] C. Tsai, "Questions that should be asked but aren't." [Online]. http://www.dfma.com/forum/2016pdf/MeekerTsaiPaper.pdf

[18] M. N. Sudin, N. S. Chin, S. A. Shamsudin, and M. A. Yusuff, "Design efficiency analysis towards product improvement for eco-friendly using DFMA method," *Open Mech. Eng. J.*, vol. 10, no. 1, pp. 173–181, Oct. 2016, doi: 10.2174/1874155X01610010173.

[19] J. C. Braga, M. L. A. C. de Castro, and M. T. D. M. Colessanti, "Method of decision making applied to the analysis of the environmental performance of a product: The case study of a Spirit fan," *Strateg. Des. Res. J.*, vol. 6, no. 2, Jan. 2014, doi: 10.4013/SDRJ.2013.62.01.

[20] M. V. A. Raju Bahubalendruni, B. B. Biswal, M. Kumar, and R. Nayak, "Influence of assembly predicate consideration on optimal assembly sequence generation," *Assem. Autom.*, vol. 35, no. 4, pp. 309–316, Sep. 2015, doi: 10.1108/AA-03-2015-022.

[21] R. Sharma, G. Singh, A. Pandey, and S. K. Sharma, "An economic inventory model comparing stock dependent and fixed demand for random machine breakdown, repair time and deterioration," *Evergreen*, vol. 10, no. 1, pp. 430–437, 2023, doi: 10.5109/6781103.
[22] M. Sharma, N. L. Jain, and J. K. Purohit, "Analysis of circular economy enablers in manufacturing context for indian industries: A electre method ranking process," *Evergreen*, vol. 10, no. 3, pp. 1156–1168, 2023, doi: 10.5109/7148437.

Chapter 7

Investigation of fault diagnosis for motor using pattern recognition method

B. K. Pavan Kumar, Yadavalli Basavaraj, N. Keerthi Kumar, C. M. Manasa, K. Lokesha and V. Venkataramana

7.1 INTRODUCTION

In the modern era, industries have well-equipped facilities to maintain the machines in a stable condition. Machines operating at higher speeds start to fail, if it's not maintained regularly, which in turn affects the health of machines and leads to major disasters in industries [1]. Traditionally, the most challenging aspect of creating rotating machine fault detection and diagnosis approaches has been extracting and quantifying the complex signal processing-based problem signals. Faults may be accurately identified utilizing previously derived attributes using a mix of data mining approaches. However, there are now difficulties in applying existing methodologies for rotating machinery failure diagnostics. Massive real-time data was collected from health monitoring systems for defect diagnostics in the age of the Internet of Things and Industrial 4.0 [2].

7.2 FAULT DIAGNOSIS OF CRITICAL MACHINE OF MOTOR

One of the issues with fault classification is the high dimensionality of rotating equipment. Data of motor machine is collected in three-dimensional positions and are analyzed. Many factors, such as load, saturation, unexpected operating conditions, electrical noise, and temperature, may all impact the fault identification and diagnosis process [3, 4]. This can result in dozens of distinct patterns that can be used to denote the vibration measurement. By studying the vibration amplitude pattern as shown in the below critical component of machine, a localized problem may be identified without affecting the other bearings in the equipment (Figure 7.1).

The reason for failure was determined by thoroughly inspecting the sources of bearing problems, such as misalignment, mechanical looseness, and so on, and removing them one by one, which causes bearing failure [5]. The details of mill fan motor are shown in Table 7.1 and identified as high vibration response.

DOI: 10.1201/9781003621102-7

Figure 7.1 Critical machine – Motor non-drive end.

Table 7.1 Specifications of mill fan motor

Type of equipment	*Motor fan details*
Location	Non-Drive End
Pressure	780 MM WC
Impeller diameter	2750 MM
Motor	Type 3-Phase Induction Motor
Rating	590Kw
Operating voltage	7KV
Full load current	71 Amp
Motor speed	990 RPM

The data measurements are collected from a variety of sources and analyzed to identify equipment failure trends and determine what maintenance is required [6, 7]. Data capture, data manipulation, status detection, health evaluation, and prognosis assessment are all activities carried out during this stage and represented in Table 7.2.

As per the above data collected with triaxial directions for different frequency, Figure 7.2 is drawn to show the highest peak in order to find the fault in the machine [8–10]. The common type of fault detection is generally categorized into two types, namely data-driven and model-based approaches. This spectrum analysis approach is effective in reducing high dimension, and the outcomes of non-overlapping in classification will make the classification process easier to diagnose.

Table 7.2 Data acquisition for motor non-drive end

Motor – Non-drive end (NDE)				
	Frequency	*Velocity*		
SL NO	*X in Hz*	*V-H*	*V-V*	*V-A*
1	20	0.01	0.04	0.01
2	40	0.01	0.05	0.05
3	60	0.07	0.12	0.21
4	80	0.04	0.08	0.33
5	100	0.31	0.35	0.31
6	120	0.01	0.1	0.1
7	140	0.01	0.05	0.05
8	160	0.01	0.03	0.08
9	180	0.01	0.08	0.05
10	200	0.01	0.01	0.06
11	220	0.01	0.01	0.02
12	240	0.01	0.01	0.01
13	260	0.01	0.02	0.03
14	280	0.01	0.01	0.02
15	300	0.02	0.03	0.01
16	320	0.01	0.05	0.05
17	340	0.01	0.02	0.02
18	360	0.01	0.01	0.07
19	380	0.02	0.01	0.05
20	400	0.01	0.01	0.03

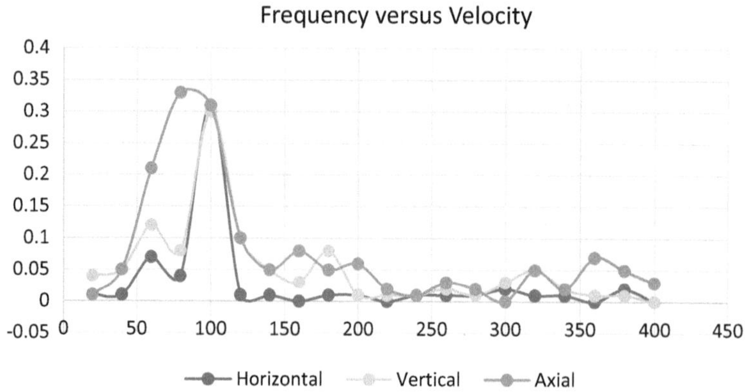

Figure 7.2 Graph shows frequency versus velocity with triaxial directions.

7.3 NEURAL NETWORK

Neural networks have been used for many decades with a lot of surveys. For the successful process of spectrums, a neural network is implemented and also well suited for pattern recognition. Effective network consists of input and hidden and output layers. In this study, we have considered 3 input layers, 10 hidden layers, and 1 output layer that satisfies the criteria for network and achieves accuracy shown in Figure 7.3. All training input is trained using a neural network to reduce error between target and training outputs [11, 12].

There are three layers to the structure of a neural network algorithm, namely input, hidden, and output layers.

7.4 ALGORITHM AND PROGRESS

It is important to understand that the algorithm does not always need labels to find similarities. When there are no helpful human labels to learn from, it uses machine learning to learn on its own, which is referred to as unsupervised learning [13]. This preserves the possibility of producing highly accurate models (Figure 7.4).

7.5 RESULTS AND DISCUSSIONS

The number of epochs isn't particularly significant. The validation and training errors are more important. Training should continue as long as it is dropping. For example, if the validation error begins to rise, this could be an indication of overfitting [14–16]. Set the number of epochs as high as possible and stop training based on error rates (Figures 7.5 and 7.6).

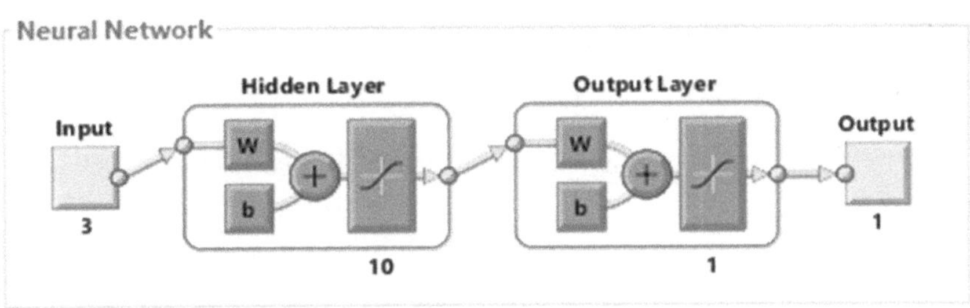

Figure 7.3 Neural network with 3 layers.

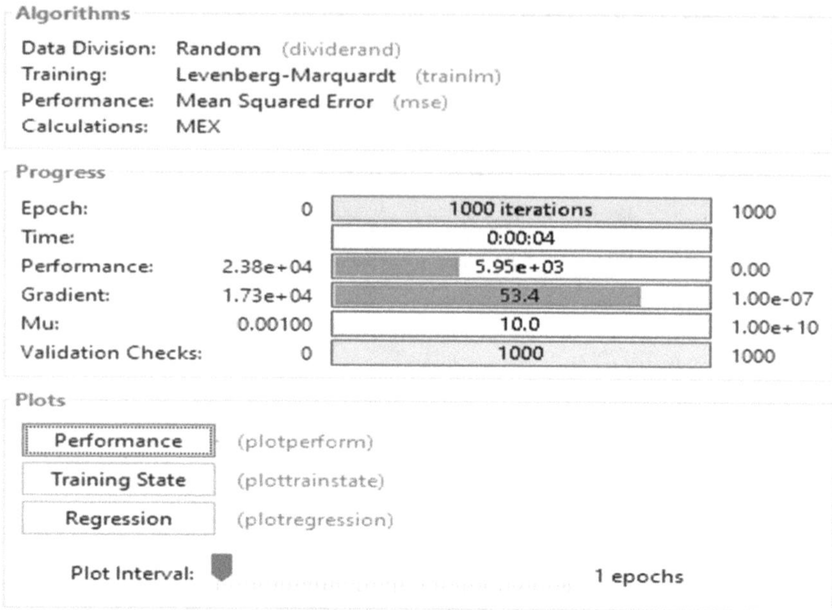

Figure 7.4 Algorithm and progress plot in values.

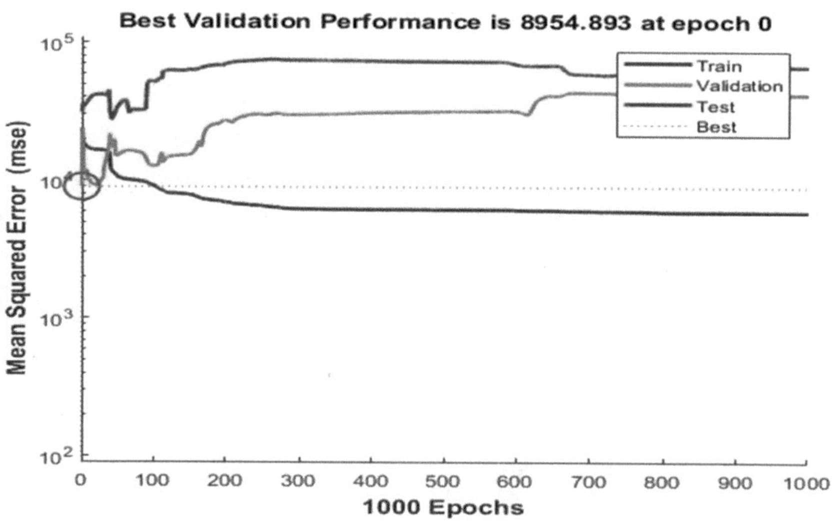

Figure 7.5 Best performance epoch value.

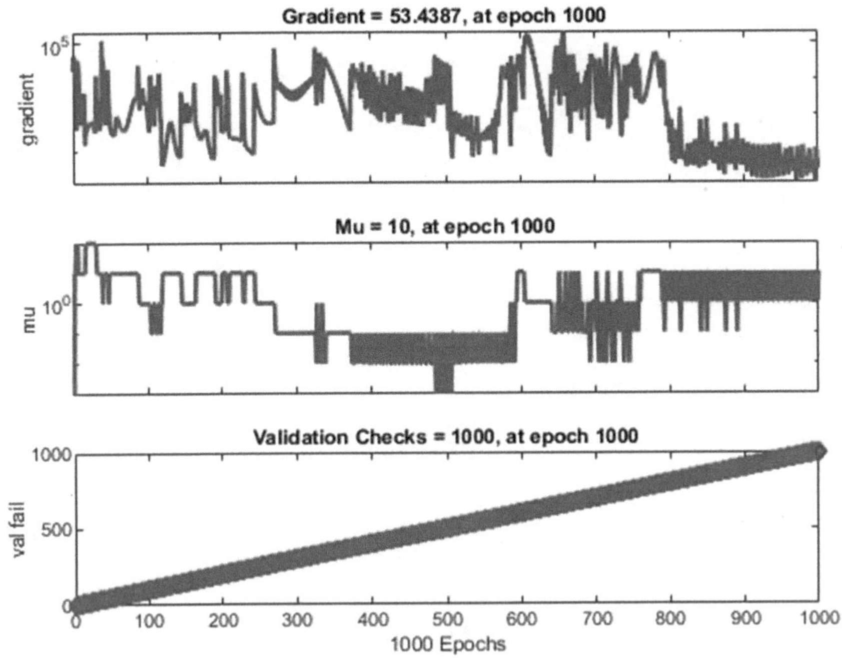

Figure 7.6 Gradient and validation checks.

7.6 VALIDATION RESULTS

Cross-validation is an important step in the development of predictive model due to various types of cross-validation methods, and the basic idea is to repeat the following process several times [17, 18].

We may now endeavor to anticipate the test set esteems and register the MSE. Since the net will produce a standardized estimate, we should downsize it to make an important correlation. Outwardly investigating the plot, we can see that the brain organization's expectations are more focused around the line than those of the straight model; an ideal arrangement with the line would recommend an MSE of 0 and consequently an optimal wonderful forecast [17–20]. Figure 7.7 shows the validation of the results using Levenberg algorithm for obtaining regression values which has a higher accuracy rate of 99%.

7.7 CONCLUSIONS

A generic methodology for detecting machinery faults using a pattern regression technique is proposed in this study. This includes collecting data, extracting features, reducing high-dimensional data, and classifying it with MPL and nearest neighbor. Although we used bearing fault diagnostics as examples,

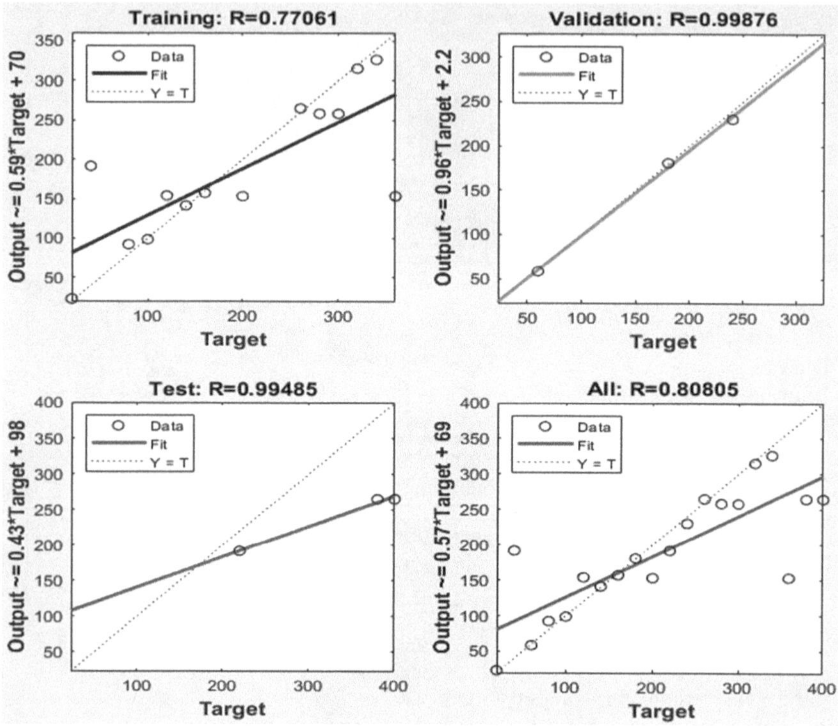

Figure 7.7 Training, test, and validation results for R-value plot.

the suggested technique can be applied to a variety of applications by simply changing the sensory signal properties. When validated using a neural network, the described approaches were found to be more than 99% accurate for rotating equipment defect identification and diagnosis. Finally, a maintenance strategy is implemented so that it is completed before the scheduled date.

REFERENCES

[1] M. Kordestani, M.F. Samadi, M. Saif, K. Khorasani, "A new fault diagnosis of multifunctional spoiler system using integrated artificial neural network and discrete wavelet transform methods", *IEEE Sens. J.* 18 (12) (2018) 4990–5001.

[2] D. Jiang, K. Wu, D. Chen, G. Tu, T. Zhou, A. Garg, L. Gao, "A probability and integrated learning-based classification algorithm for high-level human emotion recognition problems", *Measurement* 150 (2020) 107049.

[3] M. Elhoseny, K. Shankar, "Optimal bilateral filter and neural network based denoising method of medical image measurements", *Measurement* 143 (2019) 125–135.

[4] L. Gao, G. Zhang, B. Yu, J. Wang, "Wearable human motion posture capture and health monitoring based on wireless sensor networks", *Measurement* 166 (2020) 108252.

[5] G. Zhao, X. Liu, B. Zhang, Y. Liu, G. Niu, C. Hu, "A novel approach for analog circuit fault diagnosis based on deep belief network", *Measurement* 121 (2018),170–178.
[6] Y. Qin, X. Wang, J. Zou, "The optimized deep belief networks with improved logistic sigmoid units and their application in fault diagnosis for planetary gearboxes of wind turbines", *IEEE Trans. Ind. Electron.* 66 (5) (2019) 3814–3824.
[7] S. Ma, M. Chen, J. Wu, Y. Wang, B. Jia, Y. Jiang, "High-voltage circuit breaker fault diagnosis using a hybrid feature transformation approach based on random forest and stacked auto-encoder", *IEEE Trans. Ind. Electron.* (2018).
[8] F. Jia, Y. Lei, J. Lin, X. Zhou, N. Lu, "Deep neural networks: A promising tool for fault characteristic mining and intelligent diagnosis of rotating machinery with massive data", *Mech. Syst. Signal Process.* 72–73 (2016) 303–315.
[9] H. Jiang, H. Shao, "Intelligent fault diagnosis of rolling bearing using an improved recurrent neural network", *Measurement science and technology*, 29, 065–073, 2018.
[10] S. Sheng, "Wind Turbine Gearbox Condition Monitoring – Vibration Analysis", *Applied Systems Health Management Conference*, Virginia, 2011.
[11] I. M. Jamadar, S. A. I. Bellary, R. A. Kanai, A. A. Alrobaian, "Model-Based Condition Monitoring for the Detection of Failure of a Ball Bearing in a Centrifugal Pump", *Journal of Failure Analysis and Prevention, Qassim University, Buraydah, Saudi Arabia, ASM International* 2019, 11 October 2019.
[12] M. Lokesha, Manik Chandra Majumder, K. P. Ramachandran, Khalid Fathi Abdul Raheem, "Fault detection and diagnosis ingears using wavelet enveloped power spectrum and ANN", *International Journal of Research in Engineering and Technology* 02, 146–158, Sep 2013.
[13] Chunzhi Wu, Pengcheng Jiang, Chuang Ding, Fuzhou Feng, Tang Chen, "Intelligent fault diagnosis of rotating machinery based on one-dimensional convolutional neural network", *Computers in Industry* 27, 53–61, February 2019.
[14] Pankaj Gupta, M. K. Pradhan, "Fault detection analysis in rolling element bearing: A review", *International Conference of Materials Processing and Characterization (ICMPC 2016)*, Materials Today: Proceedings, 2017, pp. 2085–2094.
[15] Ana Gonzalez Muniz, Ignacio Diaz, Abel Cuadrado, " *DCNN for condition monitoring and fault detection in rotating machines and its contribution to the understanding of machine nature*", Heliyon Journal 2020, 3395–3405, 2020.
[16] James Wakiru, Liliane Pintelon, "A data mining approach for lubricant-based fault diagnosis", Journal of Emerald insights 13, 1355–1365, May 2020.
[17] Pengfei Liang, Chao Deng, Jun Wu, "Compound Fault Diagnosis of Gearboxes via Multi-label Convolutional Neural Network and Wavelet Transform", *Computers in Industry*, 1–10, 2019.
[18] Soumava Boral, Sanjay Kumar Chaturvedi, "A case based reasoning system for fault detection and isolation, A case study on complex gearboxes", *Emerald insight, Journal of Quality in Maintenance Engineering* 25(2), 2019, 213–235.
[19] Weiwei Qian, Shunming Li, Pengxing Yi, Kaicheng Zhang, "A novel transfer learning method for robust fault diagnosis of rotating machines under variable working conditions", *Journal of measurement, science direct, china*, 514–525, 2019.
[20] Fafa Chen, Yunpeng Yang, Baoping Tang, "Performance degradation prediction of mechanical equipment based on optimized multi-kernel relevant vector machine and fuzzy information granulation", *Laboratory of Hydroelectric Machinery Design & Maintenance, China*, 1–12, 2020.

Chapter 8

Application of image processing in agriculture

Dipra Mitra, Kumar Gaurav, Pallab Banerjee and Sudeshna Sani

8.1 INTRODUCTION

Deep learning techniques have been successfully used in machine vision, and their improved performance has been extended to agricultural image processing by integrating them with more conventional techniques [1]. Agricultural automation has been transformed by semantic segmentation techniques, which are now widely used for identifying pests and diseases, analysing crop cover and kind, and other tasks in order to semantically segment agricultural photos using various segmentation concepts.

In this chapter, utilising digital image processing techniques, the methodologies used to diagnose, quantify, and categorise plant diseases using digital pictures in the visible spectrum are reviewed. Figures 8.1 and 8.2 represent the infection of strawberries and the fungal plant pathogen, respectively. Although disease symptoms can manifest in various parts of a plant, this analysis only considered strategies that specifically target the observable symptoms present in leaves and stems. There were two primary motivations for undertaking this action: first, to condense the paper; and second, due to the unique attributes associated with handling roots, seeds, and fruits, which require a more comprehensive analysis.

Image segmentation techniques that expand the dataset or include multimodal data allow deep learning techniques to further enhance the segmentation capabilities. This review serves as a resource for using picture semantic segmentation in the field of agricultural information technology. In order to advance in this field, it is imperative to undertake certain tasks such as the development of picture datasets, the selection of auxiliary domains within big data, and the enhancement of the transfer learning approach. The creation of image datasets captured under authentic agricultural conditions is found to be of utmost importance in the development of agriculturally viable disease identification systems.

The identification of agricultural diseases through image recognition plays a pivotal role within the realm of intelligent agriculture. In recent years, there has been a growing application of advanced machine learning techniques, such as deep learning and transfer learning, in the field of

 DOI: 10.1201/9781003621102-8

Figure 8.1 Infection of strawberry.

Figure 8.2 Fungal plant pathogen.

agricultural disease identification [2]However, implementing these techniques still faces several noteworthy challenges. This study focuses on the analysis of deep learning and transfer learning techniques and their recent advancements in the domain of agricultural disease image recognition. According to the analysis and comparison of these two approaches, transfer learning is preferable, given the available agricultural disease datasets.

The diseased area of a leaf should be properly segmented in order to collect discriminant features. From a particular standpoint, this research presents a proposed model that is deemed efficient in diagnosing leaf diseases. The model achieves this by extracting distinguishing features from leaf images that have been segmented based on the most optimal segmentation techniques available. This study employs segmentation techniques to differentiate between leaf portions that are affected by disease and those that are in a healthy state. Subsequently, numerous identifiable characteristics are extracted from the leaf image by analysing the colour and intensity values of the individual pixels comprising the image. The multiclass support vector machine (MC-SVM) with an RBF (Radial Basis Function) Gaussian kernel is employed for the purpose of leaf illness identification and diagnosis. The proposed approach is validated using a leaf image dataset called "Plant Village-Dataset," which has been benchmarked online. The performance evaluation of the proposed model demonstrates a satisfactory outcome in accurately identifying diseases. To enhance the precision of disease detection, three widely recognised picture segmentation techniques are employed, alongside an investigation into an optimal segmentation approach. According to the experimental findings, the utilisation of the Otsu thresholding-based approach for disease segmentation demonstrates superior performance in accurately identifying leaf diseases.

Picture segmentation is the process of splitting or categorising a picture into various regions. Right now, there are many different approaches to segmenting images, ranging from simple thresholding methods to sophisticated colour segmentation methods. These components often relate to anything that people may readily divide into separate pieces and perceive. Numerous methods for segmenting photographs exist since computers are unable to recognise objects in complex ways. The segmentation procedure is built on the numeric properties of the input picture. This might be a portion, the borders, or the colour of an image. Using a genetic algorithm, we segment coloured pictures.

Detection methods for plant leaf diseases now in use are based on the characteristics of extracting algorithms. These algorithms have some restrictions on the features that can be used to identify the sick area, but they can be combined with other image processing techniques. Plant diseases can be categorised based on their symptoms. A five-step method for recognising cucumber leaves was put forth by us. These processes are preprocessing, normalisation, features extraction, features fusion, and classification. Tan-Triggs normalisation is used to normalise the dataset after Otsu's thresholding is applied during preprocessing. During the features extraction stage, texture and shape features are extracted. Additionally, expanding the instances improves various functions. Serial feature fusion and a principal component analysis technique are used to get a feature score. A support vector can be used to categorise merged features. Detection methods for plant leaf diseases now in use are based on the characteristics of extracting algorithms. These algorithms have some restrictions on the features that can be used to identify the sick area, but they can be combined with other image processing techniques. Plant diseases can be categorised based on their symptoms. A five-step method for recognising cucumber leaves was put forth by us. These processes are preprocessing, normalisation, features extraction, features fusion, and classification. Tan-Triggs normalisation is used to normalise the dataset after Otsu's thresholding is applied during preprocessing. The features extraction phase includes the extraction of texture and form features. Additionally, by increasing the instances, various functions are improved. Principal component analysis and serial feature fusion are used to create a feature score. A support vector can be used to categorise merged features.

Picture segmentation is the process of splitting or categorising a picture into various regions. Right now, there are many different approaches to segment images, ranging from simple thresholding methods to sophisticated colour segmentation methods. These components often relate to anything that people may readily divide into separate pieces and perceive. Numerous methods for segmenting photographs exist since computers are unable to recognise objects in complex ways. The segmentation procedure is built on the numeric properties of the input picture. This might be a portion, the borders, or the colour of an image. Using a genetic algorithm, we segment coloured pictures.

By using feature extraction for image data, the important portions of a picture are represented as a tiny feature vector. Previously, the task was achieved through the utilisation of specialised techniques such as feature detection, feature extraction, and feature matching methodologies. Deep learning is increasingly being utilised for the analysis of images and videos due to its ability to directly analyse raw image data without the need for preliminary feature extraction. In computer vision applications, such as image registration, object recognition and classification, and content-based picture retrieval, the selection of an appropriate image representation is crucial regardless of the chosen methodology. The topic being examined pertains to properties. The first layers of a deep network can be used to represent this intuitively or explicitly by employing some of the time-tested picture feature extraction approaches.

A variety of graphs and visuals that describe the feature extraction process are also generated; these are helpful for interpreting and comprehending the procedure. The programme is applied to a genuine parcel-based classification problem, yielding some findings and allowing for the analysis of applicability, future methodology improvement, and the usage of different types of datasets. A test of the processing speeds is also provided. This software is designed to be a dynamic tool that incorporates new data and feature extraction techniques to gradually enhance the procedures for classifying land use/land cover databases and updating agricultural databases.

For this investigation, the form characteristics that were retrieved were solidity, extension, minor axis length, and eccentricity. To locate the diseased portion of the leaf under investigation, these criteria are employed. To locate the diseased portion of the leaf under investigation, these criteria are employed. Next, it will be possible to identify how each pixel and its neighbours vary. Regarding translation, scaling, and rotation of a picture, the colour feature extraction has a unique way of showing it. The criteria utilised for colour are kurtosis, skewness, and mean by converting RGB (Red Blue Green).

8.2 IMAGE PROCESSING-BASED RECOGNITION SYSTEM

Plant disease recognition using image processing is an area of research that encompasses the application of computer vision methodologies to discern and categorise diseases in plants by analysing visual representations of their leaves, stems, or other anatomical components [5]. This approach has gained significant attention in agriculture and plant science because it offers a non-intrusive and potentially automated method for detecting diseases early, which can help farmers and researchers take timely actions to prevent or mitigate crop losses.

Here is a general overview of how image processing-based plant disease recognition works:

1. **Image Acquisition:** High-quality images of plant leaves or other relevant parts are captured using cameras or other imaging devices. These images are typically taken in controlled environments to ensure consistent lighting and background conditions.
2. **Preprocessing:** Preprocessing operations may be applied to the obtained pictures to improve their quality and eliminate noise. Operations such as resizing, cropping, colour correction, and noise reduction may be used for this.
3. **Feature Extraction:** The extraction of pertinent features is performed on pre-processed images. These features may encompass attributes such as colour information, texture patterns, shape characteristics, and other relevant factors. Feature extraction techniques could include methods like histogram analysis, texture analysis (using techniques like GLCM i.e Gray level co-occurrence matrixDL or Gabor filters), and edge detection.
4. **Classification:** Subsequently, a classification model is trained utilising the acquired features. Convolutional neural networks (CNNs), support vector machines (SVMs), random forests, and deep learning architectures are among the machine learning techniques that can be utilised for this objective. Based on the characteristics retrieved from the photos, the model learns to differentiate between healthy and unhealthy plants.
5. **Training Data:** To train the classification model, a labelled dataset is required. This dataset contains images of plants along with corresponding labels indicating whether they are healthy or diseased. Creating and annotating a diverse and representative dataset is crucial for achieving accurate results.
6. **Model Validation:** After training, the model's performance is evaluated using a separate set of images that it hasn't seen before. This helps assess its ability to generalise to new data and accurately classify healthy and diseased plants.
7. **Deployment:** Once the model achieves satisfactory accuracy, it can be deployed for real-world applications. Farmers or agricultural experts can use a mobile app or a computer software that utilises the trained model to analyse images of plants and provide disease diagnosis.

8. **Continuous Improvement:** The model can be continuously improved by periodically updating it with new data. As more labelled images become available, the model can adapt and enhance its accuracy over time.

Challenges in image processing-based plant disease recognition include variations in lighting conditions, diverse plant species, leaf shapes, and the presence of other factors that may affect the appearance of the plant. Despite these challenges, this approach holds great potential for revolutionising the way plant diseases are detected and managed in agriculture, leading to improved crop yield and reduced losses. Machine learning-based plant disease recognition:

In order to recognise and categorise illnesses in plants based on photographs, machine learning-based plant disease recognition algorithms use the capabilities of machine learning techniques. Here is a step-by-step outline of a typical machine learning-based plant disease recognition algorithm:

1. **Data Collection and Preprocessing:** Acquire a comprehensive dataset comprising a wide range of plant images, encompassing both specimens in a healthy state and those afflicted with diseases, while ensuring the inclusion of accurate labels that denote the specific type of disease. Preprocess the images by resizing, normalising, and augmenting them to ensure consistent input for the algorithm. Augmentation can involve techniques like rotation, flipping, and brightness adjustments to enhance the dataset's variability.
2. **Feature Extraction:** The pertinent characteristics are extracted from the pre-processed images. For traditional machine learning algorithms, these features could include colour histograms, texture features (such as GLCM), and other engineered representations for deep learning approaches. Feature extraction from the photos may be done automatically using CNNs.
3. **Dataset Partitioning:** The dataset is partitioned into distinct subsets, namely the training, validation, and testing sets. 7o-15-15 is the most generally used partitioning, but this can vary based on the dataset size and characteristics.
4. **Model Selection and Training:** Choose an appropriate machine learning model. For deep learning, CNN architectures (e.g., VGG, ResNet, and Inception) are commonly used due to their effectiveness in image recognition tasks. Utilise the training dataset to run the chosen model. By modifying its internal parameters during training, the model learns to associate illness labels with input photos.
5. **Hyperparameter Tuning:** Fine-tune the model's hyperparameters (learning rate, batch size, number of layers, etc.) using the validation dataset to optimise performance.
6. **Model Evaluation:** The testing dataset should be employed to evaluate the performance of the trained model. The evaluation of this can

be conducted by employing metrics such as accuracy, precision, recall, and F1-score.

7. **Deployment and Inference:** Once the model performs well, deploy it in an application or system that allows users to upload plant images for disease detection. Users submit images, and the deployed model predicts whether the plant is healthy or affected by a disease.
8. **Ongoing Improvement:** Continue to collect new data and update the model periodically to accommodate new diseases or variations in existing ones. Fine-tune the model based on real-world user feedback and additional labelled data.
9. **Interpretability and Explainability:** Consider using techniques to explain and interpret the model's decisions. Methods like feature. Understanding the reasoning behind the model's predictions may be achieved through visualisation, saliency maps, and gradient-based techniques.
10. **Post-processing and Visualisation:** Apply post-processing techniques to refine the final output, such as removing small false positives or smoothing prediction maps. Provide clear and informative visualisations to users, highlighting disease areas and potential next steps. Remember that the success of a machine learning-based plant disease recognition algorithm depends on factors such as the quality of the dataset, the choice of model architecture, hyperparameter tuning, and continuous improvement based on real-world data and feedback. Collaboration between domain experts and machine learning practitioners is essential for achieving accurate and effective disease recognition. Machine learning-based plant disease recognition algorithms leverage the power of machine learning techniques to identify and classify diseases in plants based on images. Here's a step-by-step outline of a typical machine learning-based plant disease recognition algorithm: Plant diseases are abnormal conditions that affect the health and growth of plants, resulting in various symptoms, decreased yield, and sometimes death. These diseases can be caused by pathogens such as fungi, bacteria, viruses, and nematodes, as well as by abiotic factors like environmental stressors. Plant diseases can have significant economic and ecological impacts, affecting agriculture, horticulture, forestry, and natural ecosystems. Here are some common types of plant diseases and their characteristics:

8.3 DEEP LEARNING-BASED SYSTEM

Building a recommendation system for plant disease using deep learning involves a multi-step process. Here is a high-level outline of the steps you can follow:

1. Data Collection and Preprocessing:
   Gather a comprehensive dataset of plant images that include healthy plants and various disease-infected plants. Each image should be labelled with the corresponding disease type. Preprocess the images by resizing, normalising, and augmenting them to enhance the model's performance and generalisation.
2. Data Partitioning:
   The dataset should be divided into training, validation, and testing sets. The evaluation of the performance is conducted by utilising the testing set. The validation set was utilised for the purpose of refining the hyperparameters.
3. Model Selection:
   Choose an appropriate deep learning architecture for your recommendation system. CNNs are commonly used for image-related tasks due to their ability to capture spatial hierarchies in images. You can consider popular architectures like VGG, ResNet, or EfficientNet as a starting point.
4. Transfer Learning:
   Utilise transfer learning by initialising your chosen architecture with pre-trained weights from a large dataset (e.g., ImageNet). This helps your model learn features that are generally useful for image classification tasks.
5. Model Customisation:
   Fine-tune the pre-trained model's layers on your specific plant disease dataset. You'll need to replace the last classification layer(s) with new layers that match the number of disease classes in your dataset.
6. Training:
   Proceed with training the customised model using the provided training dataset. To mitigate the risk of overfitting, it is advisable to closely monitor the loss and accuracy metrics on the validation set. To enhance the efficiency of training, it is recommended to incorporate strategies such as learning rate scheduling and early stopping.
7. Evaluation:
   Analyse the testing dataset's performance with your model. Metrics like accuracy, precision, recall, and F1-score should be used to evaluate how effectively the model is able to distinguish between different plant diseases.
8. Recommendation System Implementation:
   To create a recommendation system, you can use the trained model to classify the input plant images into different disease categories. You can then provide recommendations based on the detected disease. Additionally, you can include information about the disease, its symptoms, and potential treatments.

Table 8.1 Performance comparison of different DL (Deep Learning) architectures

Model	*Training acc (%)*	*Testing acc (%)*	*Loss*	*Epoch*	*Avg time (s/epoch)*
AlexNet	—	98.64	0.0658	50	7034
VGG	—	98.87	0.0542	49	4208
NASNet	—	93.82	—	9	—
ResNet	—	92.56	—	—	—
Deep CNN	—	96.3	—	100,000	—
Nine Layer CNN	97.87	96.46	0.2487	3000	—
CNN	99.99	97.1	—	—	—
Inception V3	98.92	98.42	0.0129	50	1027
InceptionResNetV2	99.47	99.11	0.0241	50	836
MobileNetV2	97.17	97.02	0.0921	50	565
EfficientNetB0	99.78	99.56	0.0091	50	545

9. User Interface (UI) Development:
   Develop a user-friendly interface that allows users to upload images of their plants. Display the results of the disease classification and provide recommendations for disease management.
10. Feedback Loop and Improvement:
    Collect user feedback and continuously improve your recommendation system. You can retrain the model periodically with new data to improve its accuracy and ability to identify new disease strains.
11. Scaling and Deployment:
    If your recommendation system performs well, you can deploy it to a web server or a mobile app so that users can access it easily. Ensure that the deployed system is responsive, secure, and can handle multiple requests simultaneously.

Developing a resilient recommendation system for the detection of plant diseases through the utilisation of deep learning techniques entails a multifaceted undertaking. It requires expertise in data preprocessing, model selection, training, and UI design. Collaboration with domain experts and continuous iteration based on user feedback will contribute to the success of your system. Table 8.1 presents a comparison among various deep learning architectures.

8.4 MECHANISM OF NEURAL NETWORKS IN AGRICULTURE

Neural networks have found diverse applications in agriculture, ranging from crop monitoring and disease detection to yield prediction and resource optimisation. Figure 8.3 represents a multi-layer neural network consisting

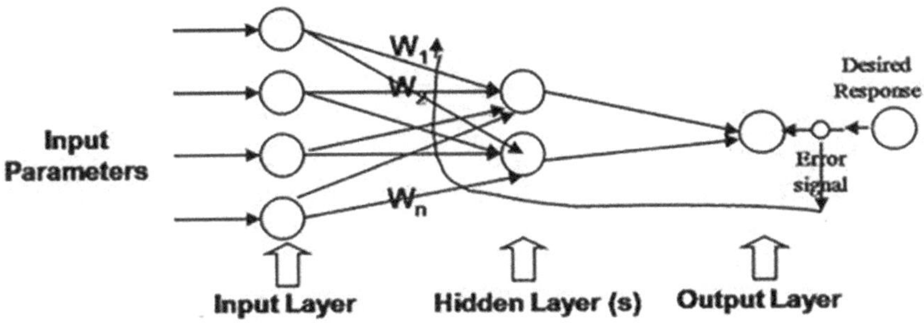

Figure 8.3 The multi-layer NN (Neural Network) consisting input layer, hidden layer, and output layer.

of input layer, hidden layer, and output layer. Here are some specific use cases of neural networks in agriculture:

1. **Crop Disease Detection and Monitoring:** Neural networks are used to analyse images of crops to detect signs of diseases, pests, and nutrient deficiencies. By training on large datasets of healthy and diseased plant images, neural networks can identify subtle visual cues that indicate crop health problems. This enables farmers to take timely actions to prevent disease spread and minimise crop losses.
2. **Yield Prediction:** Neural networks can predict crop yields based on historical data, including factors such as weather conditions, soil composition, and agricultural practices. By learning the complex relationships between these variables, neural networks provide accurate yield estimates, helping farmers plan their harvesting and marketing strategies.
3. **Weed Detection and Precision Herbicide Application:** Neural networks can distinguish between crops and weeds in images, facilitating precision herbicide application. This targeted approach reduces the need for widespread herbicide use, saving costs and minimising environmental impact.
4. **Soil Health Assessment:** Neural networks analyse soil data, such as moisture levels, nutrient content, and pH, to assess soil health. This information guides farmers in making informed decisions about fertilisation and irrigation practices.
5. **Climate and Weather Prediction:** Neural networks process historical weather data to predict future weather patterns. This information is crucial for making decisions related to planting, irrigation, and pest control.
6. **Livestock Management:** Neural networks are used to monitor the health and behaviour of livestock. They can analyse data from sensors attached to animals to predict disease outbreaks, optimise feeding schedules, and identify stress factors.

7. **Automated Harvesting:** Neural networks enable automated harvesting machinery to identify ripe fruits or crops. By recognising colour, size, and other characteristics, these networks ensure efficient and accurate harvesting.
8. **Plant Breeding and Genetic Selection:** Neural networks can analyse genetic and phenotypic data to predict the performance of different crop varieties. This assists plant breeders in selecting the most promising strains for improved yield and resilience.
9. **Market Demand Prediction:** Neural networks process market data to predict demand for specific crops. This information aids in determining what to plant and how much to produce.
10. **Supply Chain Optimisation:** Neural networks help optimise the agricultural supply chain by predicting supply and demand fluctuations, enabling more efficient distribution and reducing waste [6].
11. **Smart Irrigation:** Neural networks analyse real-time sensor data to determine optimal irrigation schedules. This conserves water resources while ensuring crops receive the right amount of moisture.
12. **Drought and Pest Outbreak Prediction:** Neural networks can predict drought conditions and pest outbreaks by analysing environmental data. This early warning system allows farmers to implement preventative measures.

Neural networks leverage their ability to recognise complex patterns and relationships within data to enhance decision-making in agriculture. By incorporating machine learning techniques, they contribute to more sustainable and efficient farming practices, supporting food security and environmental conservation.

In 1958, Rosenblatt presented the SLP (Single Layer Perceptron) mechanism and the Rosenblatt perceptron algorithm. Rosenblatt developed the perceptron rules later in 1961; these rules give the perceptron the ideal weight vector rather than the initial weight values. Additionally, it was discovered in the early 1960s that SLP may categorise the input vectors into one of two classes. The main principle of SLP is founded on an activation function, also known as the simple discriminant that changes a linear combination into a non-linear function. A version of this technique known as a "linear discriminant" transforms input variables into non-linear functions prior to creating a linear combination. Based on the linear discriminants, an SLP with changeable weight appears to be able to link the input and the output. The weights for SLP are computed using Fisher's LDA (Linear discriminant analysis), generalised linear discriminants (GLD), linear discriminant functions (LDFs), and linear separability, among other techniques. The SLP network may have one or more synthetic neurons that perceive various ideal weight vectors that can be allocated to various-

In addition, the SLP is the primary fundamental part of the multi-layer feed-forward network. It is also recognised as the most basic model for researching all non-linear multi-layer perceptrons (MLPs).

Using hyperspectral data, Monteiro et al. constructed SLP and MLP designs for visual examination of blood covering surgical fields. These designs had the capacity to learn the combinations of spectral fields' reflectance bands. Good visualisation was obtained by both systems, but SLP's output contained more noise. A straightforward and understandable verification of the viability of hyperspectral data can be offered via an SLP architecture. This application had a big impact on diagnostic hyperspectral imaging. Through this method, it may be able to do remote agricultural disease assessment. SLP architecture has not yet been used anywhere for hyperspectral data-based agricultural disease detection.

8.5 MULTI-LAYER PERCEPTRON (MLP)

In an MLP network, there are multiple hidden layers of perceptrons. In an MLP, input, output, and hidden layers are a standard set of layers. The input layer, which is the initial passive layer in an ANN (Artificial Neural Network), serves as a channel for data entry. A concealed layer makes up the second layer. The necessity of a hidden layer in a network was underlined by Paola and Schowengerdt [1] in order to improve the network's functionality and model complex situations.

The network's output signals are produced by the output layer, which is the lowest layer. As the SLP is no longer practically viable, the MLP is best suited for processing hyperspectral data, particularly in the context of non-destructive sickness diagnosis for high performance classification [1, 2]. MLP design was utilised by Moshou et al. [3] to identify yellow rot in wheat crops. The MLP architecture was built with two neurons in the output layer, one for a healthy crop and the other for a damaged crop, an input layer with neurons equal to the number of processed spectral bands, and a hidden layer with neurons varying in number from 5 to 25. The photos in the wheat field were taken using a portable spectrograph (460–900 nm). Four excellent spectral bands were chosen for this work. Different numbers of neurons were evaluated, and the most effective neurons were chosen for the final MLP architecture. The MLP design yielded a classification accuracy of over 99% for ill plants and over 98% for healthy plants [4].

Recent studies have shown that the MLP network is quite useful. The majority of MLP networks use backpropagation techniques for training. MLP is a fairly common choice among researchers as a result [5]. Backpropagation methods in MPL's supervised learning paradigm decrease errors between intended outputs and estimated outputs obtained from inputs and network learning [6].

8.6 RECENT DEVELOPMENTS IN THE APPLICATION OF IMAGE PROCESSING IN AGRICULTURE

The field of image processing in agriculture is experiencing rapid advancements, driven by a substantial body of research aimed at addressing various aspects of agricultural practices. In this analysis, we will explore several notable contributions in this field, which collectively offer valuable perspectives on the present condition and prospective advancements of image processing applications in the agricultural sector.

1. **Disease Detection and Crop Health:** Prominent research endeavours, such as the study conducted by Janarthan et al. in 2022 titled "Deep Residual Learning for Small-Footprint Crop Disease Classification with Limited Training Samples," [7] and the comprehensive survey conducted by Ferentinos in 2018, have delved into the utilisation of deep learning and machine learning methodologies for the purpose of disease identification [8]. These studies demonstrate the effectiveness of utilising a small amount of training data for the purpose of classifying diseases, thus enabling the utilisation of technology by farmers operating on a smaller scale.
2. **Prediction of crop yield:** There is a growing momentum in the utilisation of deep learning techniques for the purpose of enhancing crop yield predictions. Studies such as the one conducted by Rangrajan Arvind et al. in 2020, which explores the application of deep neural networks [9], and the comprehensive review on remote sensing and machine learning techniques conducted by Maxwell et al. in 2020 have laid the foundation for improved yield predictions that are based on accurate data analysis [10].
3. **Weed Detection:** Extensive research has been conducted on the issue of weed detection, as evidenced by the comprehensive examination conducted by Reiser et al. in 2019 [11] and the scholarly contribution made by Barbedo in 2019 [12]. These scholarly articles explore the diverse range of computer vision methodologies that can be utilised for the automated identification and spatial representation of weeds.
4. **Livestock Monitoring:** The role of sensors and computer vision in precision livestock farming has been highlighted in studies conducted by Zurnawita (2023) [13]. These works have played a significant role in advancing innovations in the monitoring of livestock behaviour and health.

8.7 EVALUATION OF QUALITY IN AGRICULTURAL YIELD

The significance of image processing in the context of automated quality assessment also deserves attention. Scholarly works such as the review conducted by Kavdir et al. in 2018 on fruit inspection [14] and the review conducted by Moghimi et al. in 2022 on quality assurance for fruits and vegetables offer comprehensive analyses of contemporary methodologies in these domains [15].

Figure 8.4 Sample image of each class.

These studies demonstrate the diversity of image processing applications in agriculture, ranging from disease detection to yield prediction and livestock monitoring. As the field continues to evolve, the integration of advanced machine learning techniques and sensor technologies is likely to play a significant role in shaping the future of smart and efficient agriculture. In particular, the emergence of CNNs has introduced novel opportunities for conducting intricate and precise analyses. In the subsequent section, we will examine the transformative impact of CNNs on diverse facets of agricultural image processing. This advancement presents a novel framework for addressing the most critical challenges faced by the agricultural industry. Figure 8.4 represents the images of late blight, early blight, and healthy leaves.

8.8 CONVOLUTIONAL NEURAL NETWORKS

CNNs are one of the types of artificial neural networks, and they are commonly used for classification, image processing, segmentation tasks, etc. [16]. Convolution is the technique of applying a filter to a picture to draw out certain important elements. Knowing that the picture in Figure 8.5 is nothing more than a matrix having a single numerical value, denoted by the

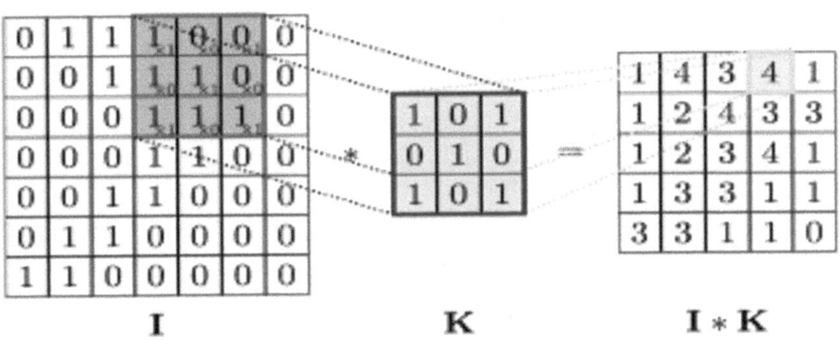

Figure 8.5 Convolution using the filter K over the image.

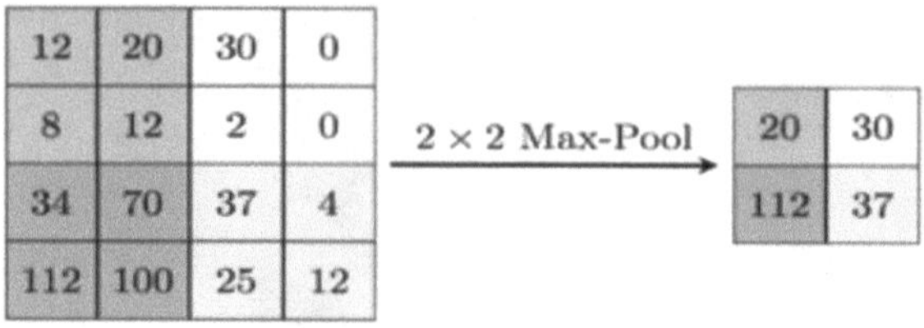

Figure 8.6 The max-pooling from the 2*2 window and lowering the dimension from 4*4 to 2*2 are shown.

letter I, is important. With the help of these filters, which are convolved over the input picture and are represented by the letter K in Figure 8.5, the key material or characteristics are subsequently learnt at different stages.

Convolutional, packing, fully linked, and normalising layers make up the CNN's hidden layer, which serves as an input to the activation function. Convolutional, packing, fully linked, and normalising layers make up the CNN's hidden layer, which serves as an input to the activation function. Each has a SoftMax activation function. Pooling is used to down sample data or to lower the dimensionality of the input representation, whereas convolution has already been described. The three different forms of piling are minimum, maximum, and average. Min pools and max pools, as indicated by their names, extract, respectively, the least value from a matrix of a certain dimension and the largest value from the matrix. Figure 8.6 below displays an example of maximum pooling using the 2*2 window.

Taking the average of all the values in that matrix is what is referred to as average pooling. Fully linked layers signify that all the neurons in the layer below will receive input from the weights from the layer above. The normalisation layer is used to maintain the mean activation near to zero while normalising the activations of the preceding layer [17].

8.9 HISTOGRAM EQUALISATION

The technique of histogram equalisation is one used for improving images. The picture intensities are distributed using this technique (Figure 8.7). This distribution raises the contrast in the affected areas from low contrast to high contrast. To enhance the image's readability, visibility, and quality, histogram equalisation is utilised. A uniform histogram is produced in the output image by histogram equalisation.

The intensity levels of the photos are distributed using the cumulative distribution function. This function determines the transformation T that converts the input image's grayscale values (r) to its transformed values (s).

$$s = T(r)o <= r <= 1$$

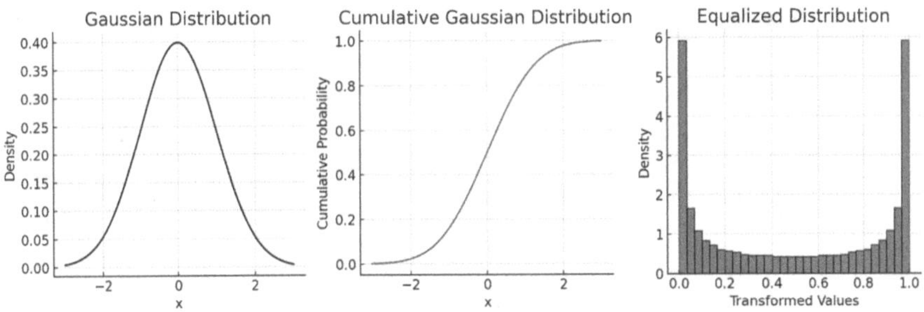

Figure 8.7 Gaussian distribution.

The following conditions are met by the transformation function T(r):

- For o = r = 1, o = T(r) = 1.
- T(r) has a single value and increases monotonically.

The mapping function y = f(x) transfers the values x to y, which are also in the same range, if the pixel values are in the range of o to x to 1. The output image's range dy is transferred to the input image's interval dx. As a result, P(y) dy = p.

To equalise the histogram's output image, constant 1 is required.

Based on Gaussian distribution, some formula has been utilised P(y) = Probability of y used

$$p(y) = 1$$

$$\text{So}, o 1 p(y) dy = o$$

$$1 dy = 1$$

As a result, we have dy=p(x)dx.

Using the integration for histogram equalisation, we compute the mapping function y = f(x):

$$p(u) du = P(x) - P(o) \text{and} y = f(x) = o$$

The input image's cumulative distribution is P(x) in this case.

$$P(o) = o \text{ and } P(x) = o x p(u) du$$

The following can be used to explain this mapping function logically:

- If dy is wide and p(y) is low, then p(x) has a steep slope and p(x) is high.
- If dy is narrow and p(x) is low, p(x) will have a shallow slope and p(y) will be high.

When dealing with discrete values, the continuous mapping function is as follows: where the input x accepts L discrete values, o=xL.

8.10 IMPROVED HISTOGRAM EQUALISATION

Several methods for enhancing the image histogram were provided. Brightness preserving bi-histogram equalisation (BBHE), bi-histogram equalisation with a plateau level (BHEPL), dualistic sub image histogram equalisation (DSIHE), quadrant dynamic histogram equalisation (QDHE), and others are popular techniques. However, a high contrast picture is necessary in order to extract the key components that will enable us to identify plant leaf disease. Therefore, we provide a better histogram equalisation approach for picture improvement. Assuming that the grey level values of the input plant leaf image I, which is represented by I, fall within the range [o, G-1], the probability density function may be characterised as follows:

$$PDF(r_k) = {}^{n_k}\!/_{N} \quad k = 0,1,2,3\ldots\ldots,G-1 \tag{8.1}$$

where rk is the Kth grey level and nk is the number of pixels in the picture. This allows for the calculation of the cumulative distribution function.

$$CDF(r_k) = \sum_{i=0}^{i=k} PDF(r_i) \text{where} k = 0,1,\ldots\ldots\ldots L-1 \text{and} 0 \leq CDF(r_k) \leq 1 \tag{8.2}$$

The image's grey level can be written as: Conventional histogram equalisation algorithms use CDF computation from Eq. (2).

$$S_k = CDF(r_k)X(G-1) \tag{8.3}$$

As seen below, the probability density function may be used to calculate changes in an image's grey level.

$$\Delta S_k = (G-1)\ XPDF(r_k) \tag{8.4}$$

However, there are certain difficult problems that CDF computation may run into, like quantisation, which hinders the performance of image quality enhancement. We present a new formula for PDF computation to address this problem. The new PDF feature is available as

$$PDF_{new} = \begin{cases} PDF_{max}\ PDF(k) = PDF(r_k) \\ \left(\dfrac{PDF(k)}{PDF_{max}}\right) X\, PDF_{max}, 0 < PDF(k) < PDF_{max} \\ 0, PDF(k) = 0 \end{cases} \tag{8.5}$$

1. **Feature extraction:** We will offer a model for heterogeneous feature extraction in the following stage of the project. We concentrate on creating a combined feature extraction method that computes and combines form, texture, and colour characteristics [18].
2. **Texture feature:** By taking the product of the sinusoid function and the Gaussian kernel, we use the Gabor filtering computation model to produce the texture feature model. Let g be a 2D Gaussian curve that has the following expression

$$g(x, y, \sigma) = e^{\left(\frac{x^2+y^2}{2\sigma}\right)} \cdot \frac{1}{2\pi\sigma^2} \tag{8.6}$$

   Once these components have been identified, the GLCM calculation scheme is performed to the hypothetical components in the following step to extract all the image's information. Table 8.2 displays the normalised co-occurrence matrix features. The distribution of GLCM features is based on the angle and distance relationship between pixels and is thought of as a matrix. Image pixels often vary at every place, and this variation can be exploited to identify different texture information using GLCM. This matrix can be used to compute several features using GLCM, including features related to visual texture, statistics, information theory, and information correlation measurement.
3. **Shape feature:** We concentrate on obtaining shape features using the curvelet transform and invariant moments after extracting texture features. Curvelet transform: A variety of computer vision applications frequently use curvelet transforms. Ridgelet analysis is used throughout this modification to extract data from the source image. In this regard, also provided is a fast discrete curvelet transform (FDCT) approach that uses fast Fourier transform and wrapping, two distinct types of functions. The wrapping function, which can evaluate different subbands at different scales and be used to extract orientation and position information in the examined frequency domain, is used in this work.

8.11 COMBINED COLOUR AND TEXTURE FEATURES

The integrated scheme for computing colour and texture features is shown in the following stage.

Images of plant leaves typically have an uneven texture, which complicates the identification of texture pattern [19]. Furthermore, chromatic information loss in traditional approaches prevents them from delivering the important texture feature.

In this chapter, we describe a novel method in which the input image is covered by a circular window that slides over it. This window aids in figuring out how regularly spaced out the pixel data is. A coloured texture feature extraction window is made in this stage to lessen the computational complexity by treating the 2D input image as a one-dimensional projection space.

1. Crop Monitoring and Management
   Disease Detection: Image processing helps identify diseases, pests, and nutrient deficiencies in plants by analysing visual symptoms [20]. Early detection allows for timely interventions, reducing crop losses [21].
   Weed Detection and Management: Images of fields can be processed to differentiate between crops and weeds, enabling targeted herbicide application and reducing chemical use [22].
   Nutrient Status Assessment: Image analysis can assess nutrient deficiencies by analysing leaf colour and structure, guiding precision nutrient application [23].
2. Precision Agriculture
   Variable Rate Application: Images are used to create prescription maps for variable rate application of inputs like fertilisers, pesticides, and water, optimising resource use and reducing environmental impact [24].
   Yield Mapping: Crop yield is estimated by analysing images and correlating them with crop health, helping farmers understand spatial yield variations [25].
3. Remote Sensing and Monitoring
   Satellite and Drone Imaging: Remote sensing platforms provide multispectral and hyperspectral data for monitoring large areas, tracking crop health, and assessing environmental conditions [24].

Table 8.2 GLCM features

Feature	*Mathematical formula*
Entropy	$\sum_i \sum_j p(i,j)\log p(i,j)$
Contrast	$\sum_i \sum_j (i-j)^2 p(l,j)$
Energy	$\sum_i \sum_j p^2(l,j)$
Homogeneity	$\sum_i \sum_j p(l,j)/i+i-j$

Drought and Stress Monitoring: Images help assess plant stress caused by factors like water scarcity or disease, aiding timely irrigation decisions [26].

4. Harvesting and Sorting
   Ripeness Detection: Image processing can determine fruit or vegetable ripeness based on colour, shape, and size, optimising harvest timing and quality [27].
   Automated Sorting: Cameras and image analysis systems sort harvested produce based on quality attributes, enhancing efficiency and reducing waste.
5. Plant Breeding and Genetics
   Phenotyping: Image-based phenotyping captures plant traits like size, shape, and colour, aiding in genetic research, trait selection, and breeding programs.
   Genomic Selection: Image data contributes to genomic prediction models, improving the accuracy of selecting desirable plant traits [28, 29].
6. Climate and Environmental Monitoring
   Crop Stress Assessment: Image analysis helps assess the impact of climate stressors (e.g., heat, cold, wind) on crops, facilitating adaptation strategies.
   Biodiversity Monitoring: Images contribute to tracking changes in plant biodiversity, habitat mapping, and conservation efforts [30].
7. Livestock Management
   Animal Health Monitoring: Image processing techniques, such as thermal imaging, assist in monitoring livestock health, detecting diseases, and managing welfare [31].
8. Post-Harvest Quality Control
   Quality Assessment: Images are used to assess the quality and defects of harvested produce, enabling better decision-making for distribution and marketing [32].
9. Education and Extension
   Farmers' Training: Image-based resources, including tutorials and guides, aid in educating farmers about various agricultural practices and challenges.
10. Economic Analysis
   Yield Estimation: Image data contributes to estimating crop yields, which is essential for economic planning and decision-making. Image processing continues to evolve, incorporating advanced technologies like artificial intelligence and machine learning. These developments promise even greater capabilities for addressing agricultural challenges and promoting sustainable practices.

8.12 CONCLUSION

Image processing has become a major force in the realm of agriculture, successfully merging the two hitherto separate fields of farming and cutting-edge technology into a single, data-driven discipline. A wide range of applications, including weed and pest identification, improved irrigation management, crop health surveillance, and harvest planning optimisation, were covered in the chapter. In agriculture, the application of image processing techniques makes it easier to transform raw visual input into insightful, factual information. As a result, farmers and other agricultural specialists are better equipped to make decisions that are tailored to their unique requirements and objectives. By spotting problems like agricultural diseases and pests early on, early diagnosis and treatment capabilities of technology enable targeted and timely actions that reduce potential losses. The previously mentioned technology advantage seamlessly meshes with precision agriculture, facilitating the economical use of resources like pesticides, fertilisers, and water. Detection and harvesting of weeds are among the jobs where this integration eventually improves operational efficiency. Predictive assessments are another feature of the system that helps with crop rotation planning and production forecasts in the future. The overriding goals of fostering environmental sustainability and guaranteeing food security are intimately related to the incorporation of technology in the agriculture sector. Moreover, laying the groundwork for intelligent and flexible agricultural systems requires incorporating other cutting-edge technologies like machine learning, artificial intelligence (AI), and the Internet of Things (IoT). However, there are still challenges to be solved with regard to data quality, privacy issues, and the computational resources needed, all of which call for cautious application and niche expertise. As we navigate this critical juncture in agricultural development, it is evident that combining expertise in technology, data science, and agriculture will be essential to maximising image processing's potential. This will become increasingly relevant as we move on to discuss more advanced applications and technologies in the upcoming chapters, such as image processing-based recognition systems, machine learning-based plant disease detection systems, and deep learning-based systems.

REFERENCE

[1] Mohanty, S. P., Hughes, D. P., and Salathe, M., "Using deep learning for image-based plant disease detection," *Frontiers in Plant Science*, vol. 7, p. 1419, 2016.

[2] Fenu, G. and Malloci, F. M., "Forecasting plant and crop disease: an explorative study on current algorithms," *Big Data and Cognitive Computing*, vol. 5, no. 1, p. 2, 2021.

[3] Harvey, C. A., Rakotobe, Z. L., Rao, N. et al., "Extreme vulnerability of smallholder farmers to agricultural risks and climate change in Madagascar," *Philosophical Transactions of the Royal Society B: Biological Sciences*, vol. 369, no. 1639, Article ID 2o13oo89, 2014.

[4] Fenu, G. and Malloci, F. M., "Lands DSS: a decision support system for forecasting crop disease in Southern Sardinia," *International Journal of Decision Support System Technology*, vol. 13, no. 1, pp. 21–33, 2021.

[5] Fenu, G. and Malloci, F. M., "Artificial intelligence technique in crop disease forecasting: a case study on potato late blight prediction," in *Smart Innovation Systems and Technologies*, pp. 79–89, Springer, Singapore, 2020.
[6] Goyal, Ankur & Kanyal, Hoshiyar & Sharma, Bhasker. (2023). Analysis of IoT and blockchain technology for agricultural food supply chain transactions. *International Journal on Recent and Innovation Trends in Computing and Communication*. 11. 234–241. 10.17762/ijritcc.v11i3.6342.
[7] Janarthan, S., Thuseethan, S., Rajasegarar, S., & Yearwood, J. (2022). P2OP—Plant Pathology on Palms: A deep learning-based mobile solution for in-field plant disease detection. *Computers and Electronics in Agriculture*, 202, 107371.
[8] Ferentinos, K. P. (2018). Deep learning models for plant disease detection and diagnosis. *Computers and electronics in agriculture*, 145, 311–318.
[9] Rangarajan Aravind, K., & Raja, P. (2020). Automated disease classification in (Selected) agricultural crops using transfer learning. *Automatika: časopis za automatiku, mjerenje, elektroniku, računarstvo i komunikacije*, 61(2), 260–272.
[10] Maxwell, A. E., Warner, T. A., & Fang, F. (2018). Implementation of machine-learning classification in remote sensing: An applied review. *International journal of remote sensing*, 39(9), 2784–2817.
[11] Reiser, D., Sehsah, E. S., Bumann, O., Morhard, J., & Griepentrog, H. W. (2019). Development of an autonomous electric robot implement for intra-row weeding in vineyards. *Agriculture*, 9(1), 18.
[12] Barbedo, J. G. A. (2019). A review on the use of unmanned aerial vehicles and imaging sensors for monitoring and assessing plant stresses. *Drones*, 3(2), 40.
[13] Zurnawita, Z., Prabowo, C., Kurnia, R., & Elfitri, I. (2023). A review of Image Processing Technique for Monitoring The Growth and Health of Cows. *JITCE (Journal of Information Technology and Computer Engineering)*, 7(01), 8–18.
[14] Kavdır, İ., Burak Büyükcan, M., & Kurtulmuş, F. (2018). Classification of olives using FT-NIR spectroscopy, neural networks and statistical classifiers. *Journal of Food Measurement and Characterization*, 12, 2493–2502.
[15] Moghimi, A., Sazgarnia, A., & Aghkhani, M. H. (2022). Spectral Feature Selection from the Hyperspectral Dataset to Identify Pistachio Leaves Infected by Psylla. *Journal of Agricultural Machinery*, 12(2), 159–167.
[16] Liu, B., Zhang, Y., He, D., and Li, Y., "Identification of apple leaf diseases based on deep convolutional neural networks," *Symmetry*, vol. 1o, no. 1, p. 11, 2018.
[17] Ahmad, I., Hamid, M., Yousaf, S., Shah, S. T., and Ahmad, M. O., "Optimizing pretrained convolutional neural networks for tomato leaf disease detection," *Complexity*, vol. 2020, 8812019, 2020.
[18] Yan, Q., Yang, B., Wang, W., Wang, B., Chen, P., and Zhang, J., "Apple leaf diseases recognition based on an improved Table 4: Comparison of the training time. *Architecture Training time* p. 3535, 2020.
[19] Mitra, D., Goyal, A., Gupta, S., Kanyal, H. S., Kaushik, S., & Kumar, K. Automated tomato leaf disease detection technique using deep learning. *Journal of Theoretical and Applied Information Technology*, 101(14), 2023.
[20] Mitra, D., Gupta, S., Srivastava, D., & Sani, S. Plant disease identification using convolution neural networks. In *Intelligent Internet of Things for Smart Healthcare Systems* (pp. 203–215), 2023.
[21] Kranz, J. Measuring plant disease. In Experimental techniques in plant disease epidemiology, pp. 35–50, Springer, 1988.

[22] Mitra, D. and Gupta, S., "Plant Disease Identification and its Solution using Machine Learning," *2022 3rd International Conference on Intelligent Engineering and Management (ICIEM)*, London, United Kingdom, 2022, pp. 152–157, doi: 1o.11o9/ICIEM54221.2o22.9853136.

[23] Mitra, D., & Gupta, S. Design and Implementation of Alex Net-Honey Badger Fusion Algorithm for Feature Selection and Classification Using Multiclass Support Vector Machine Classifier to Recognize Plant Leaf Disease. International Journal of Intelligent Systems and Applications in *Engineering*, 11(3), 7o4–711, 2023.

[24] Sishodia, R. P., Ray, R. L., & Singh, S. K. Applications of remote sensing in precision agriculture: A review. *Remote Sensing*, 12(19), 3136, 2020.

[25] Lowenberg-DeBoer, J., & Erickson, B. Setting the record straight on precision agriculture adoption. *Agronomy Journal*, 111(4), 1552–1569, 2019.

[26] Wen, W., Timmermans, J., Chen, Q., & van Bodegom, P. M. A review of remote sensing challenges for food security with respect to salinity and drought threats. *Remote Sensing*, 13(1), 6, 2020.

[27] Ko, K., Jang, I., Choi, J. H., Lim, J. H., & Lee, D. U. Stochastic decision fusion of convolutional neural networks for tomato ripeness detection in agricultural sorting systems. *Sensors*, 21(3), 917, 2021.

[28] Goyal, A., Rathore, L., Kumar, S A Survey on Solution of Imbalanced Data Classification Problem Using SMOTE and Extreme Learning Machine. *Lecture Notes in Networks and Systems*, vol 204. Springer, Singapore, 2021. https://doi.org/10.1007.

[29] Shakoor, N., Northrup, D., Murray, S., & Mockler, T. C. Big data driven agriculture: big data analytics in plant breeding, genomics, and the use of remote sensing technologies to advance crop productivity. *The Plant Phenome Journal*, 2(1), 1–8, 2019.

[30] Scherber, C., Beduschi, T., & Tscharntke, T. Novel approaches to sampling pollinators in whole landscapes: a lesson for landscape-wide biodiversity monitoring. *Landscape Ecology*, 34, 1057–1067, 2019.

[31] Garcia, R., Aguilar, J., Toro, M., Pinto, A., & Rodriguez, P. A systematic literature review on the use of machine learning in precision livestock farming. *Computers and Electronics in Agriculture*, 179, 105826, 2020.

[32] Valdez, P. Apple defect detection using deep learning based object detection for better post harvest handling. 2020. arXiv preprint arXiv:2005.06089.

Chapter 9

What happens when people join online brand communities and talk to each other

A netnographic perspective

Uttam Chakraborty and Santosh Kumar Biswal

9.1 INTRODUCTION

The total number of Internet subscribers in India has touched the mark of 751.5 million (Kepios, 2024). Digital consumption is proliferating day by day and is influencing the usage of social media in India. Facebook has the highest number of active users compared to other social media platforms in India. India has over 378.05 million active users of Facebook (Statista, 2024). India's consumer digital economy is expected to grow to a $1 trillion market by 2030, up from $537.5 billion in 2020, driven by strong adoption of online services such as e-commerce and edtech in the country (IBEF, 2024).

Muniz Jr (2001) defined brand communities as "a specialized, non-geographically bound community, based on a structured set of social relations among admirers of a brand (423)." They developed three types of markers – consciousness of kind, shared rituals and traditions, and moral responsibility – which ensure the idealness of online social media brand community. Participation can be defined as the degree to which consumers are engaged in reading, posting, and commenting in the communities which are essential for the survival of online communities. There are three types of participations, namely, informational, actionable, and attitudinal that the consumers practice in online social media (Kamboj, 2020; Labrecque et al., 2020). This study follows the uses and gratification theory (UGT) to understand the consumers' participation in social media and the gratification because of consumer participation. Therefore, the research question of the study:

9.1.1 Research question

Does consumer social participation, namely informational, attitudinal, and actionable, are present in consumer electronics brand communities?

9.1.2 Objective of the study

To analyze the evidences of consumer social participation, namely, informational, attitudinal, and actionable, are present in consumer electronics brand communities.

DOI: 10.1201/9781003621102-9

To achieve the objective, the present study follows the netnography technique because of its set of research practices in terms of data collection and analysis. The term "netnography" is a combination of the Internet or network and ethnography. An ethnographic study emphasizes on a culture-sharing group in order to find the collective arrangements of beliefs and behaviors among the participants of a certain group (Creswell and Poth, 2016). It involves a humanly affair and is immensely connected with the group of people in a particular community in a given period of time. Interestingly, netnography is an interpretive research method which resorts to participant observation techniques for probing the extent of interactions and experiences manifested through digital communications (Kozinets, 1998). The basic difference between netnography and other qualitative techniques like focused group interview, participant's interview, and qualitative content analysis is the researcher has to ask an open-ended question or raise certain kinds of issues so that the participants can discuss on any given issue. Moreover, in the context of quantitative technique, the researcher has to ask a certain amount of close-ended questions to the participants. Moreover, quantitative techniques as well other qualitative techniques except netnography, the researcher's data collection depends upon the respondents' responses. Therefore, the researcher has to trust on respondents' responses without having any sort of evidences. The prior studies measured consumer social participation through quantitative methods (Survey) (Kamboj, 2020; Nadeem et al., 2021). There is an impending need to understand consumer social participation with the technique which does the analysis from the evidences, not from the responses from the respondents.

9.2 LITERATURE REVIEW

9.2.1 Uses and gratification theory (UGT)

First, UGT was followed in the year 1944 to understand the users' choice of a certain type of media. In 1954, it was utilized as an extension of the needs and motivation theory (Katz, Blumler, and Gurevitch, 1973). In 1964, the theory was employed to examine people's intentions to view certain television programs. However, in the era of 1980s, the theory was used to probe people's perspectives of mass media (Kamboj, 2020). Consumers use social media to fulfill their social and emotional needs when they share opinion on products or services in online social media. Further, they enjoy other people's attention toward that information. Internet gratifies people in various ways. Through Internet, people gratify them by means of opinion leader and opinion seeker, easy access of information, enjoyment, and entertainment (Lim and Kumar, 2019). Facebook gratifies people in terms of interactions with or knowing about others, opinion giving, and opinion seeking (Kim and Kim, 2019).

Prior researchers have followed UGT in various contexts. For example, social media and voter relationship (Abid and Harrigan, 2020), mobile social network sites (Kim, Lee, and Contractor, 2019), social media technology (Kamboj, 2020), virtual goods (Kaur et al., 2020), food delivery apps (Ray et al., 2019), mobile shopping (Huang and Zhou, 2018), regrets in social networking sites (Islam, Mäntymäki, and Kefi, 2020), brand engagement (Florenthal, 2019), consumer brand engagement (Jayasingh, 2019), users' content contribution behaviors on microblogs (Liu, Min, and Han, 2020), photo tagging (Dhir, Chen, and Chen, 2017), and Instagram (Kim and Kim, 2019).

9.2.2 UGT and online brand community

As community members are like-minded people, they share common behavior, rituals, and traditions (Kaur et al., 2020). Further, Muniz Jr (2001) also argued that a perfect brand virtual community must retain the community markers – consciousness of kind, shared rituals and tradition, and moral responsibility (Luo et al., 2015).

Within a brand community, members' sense of belongingness among each other is very high which is considered as "consciousness of kind". Consciousness of kind gratifies the members by uniting all the members of the brand community in which individual identity no longer persists. Moreover, there is collectiveness in terms of identity where members use "we" instead of "I" when they deliberate within the brand community. This collectiveness has no geographical boundaries.

In the brand community, members share similar kinds of rituals and traditions. It also gratifies their unity. Their unity in the community empowers their expression of opinion. Members of the brand community show a higher sense of moral responsibility among themselves. They are always ready to help and suggest each other in brand-related issues which ultimately gratify their unity.

9.2.3 UGT and participation in online brand community

This chapter analyzed gratifications received by the individuals being an active member of the brand community. In the brand community, individuals actively participate by reading messages, posting responses, sharing new information, and even moderating community members' discussions among community members. Following this trend, previous literature argues that the types of consumer social participations observed in online social media are informational, actionable, and attitudinal (Kamboj, 2020).

Informational participation is when consumers share general or new information about the brand in their respective communities. These kinds of activities fulfill their needs as an opinion leader. Virtually, they become an important part of community in which can guide other members on the utility of that particular brand. Therefore, it is their sense of moral responsibility that they feel for the betterment of other community members. Further, when individuals reply to others' posts or evaluate others' performance, their psychological tendency is described as actionable participation. Actionable participation focuses on the interaction among community members. Therefore, this kind of engagement behavior gratifies their emotional bonding, trust, commitment, connection, and self-empowerment (Kamboj, 2020).

According to Brodie et al. (2013), consumer engagement can be referred as, "The level of a customer's motivational, brand-related and context-dependent state of mind characterized by specific levels of cognitive, emotional and behavioral activity in brand interactions (106)". Moreover, when members of the community share their positive opinions or joy or positive emotions about the brand, this can be treated as attitudinal participation of the brand community members. When consumers share their happiness, satisfaction, and positive vibes, this can improve the members' loyalty to the brand. On the other hand, if the members share their negative opinions or experiences, this can degrade their loyalty or acceptance of the brand.

9.2.4 Netnography

Netnography, the technique used for the current study, was proposed in the year 1995. This technique focuses on understanding social and cultural interactions in digital communication. It has the uniqueness of research practices, including data collection, analysis, and participation observation. It is an online ethnography technique. There are basic differences between ethnography and netnography. In case of an ethnographic study, the researcher is actively available in a physical space in a community within a time frame. On the other hand, digital ethnography is determined based on the discursive interactions among members of a given community (Kozinets, 2010).

Netnography analyzes the feelings and emotions of consumers on a brand (Chakraborty and Bhat, 2019) which remains essential to probe consumer behavior in the domain of brand communication (Loureiro, Correia, and Guerreiro, 2019). It is used to examine the layers of social interactions in digital platforms. Kozinets laid certain entrée which are the qualifiers to conduct netnography. These entrées are relevant, active, interactive, substantial, heterogeneous, and data-rich.

Nowadays, netnography is a tool which is popularly used not only in marketing and consumer research but also in the field of sociology, psychology, anthropology, geography, education, hospitality, and tourism. However, online ethnography is predominantly being used in sociology, communication, and anthropology. This qualitative approach to research "uses computer mediated communications as a source of data to arrive at the ethnographic understanding and representation of a cultural or communal phenomenon" (Kozinets, 2010, p. 60). The extant studies have followed netnogrpahy in various contexts. For example, tourism (Tavakoli and

Wijesinghe, 2019; Kozinets and Gretzel, 2024), social media (te Lintelo, Hernandez, and Lakshman, 2024), higher education (Eaton and Pasquini, 2020), transgender community (Linsenmeyer and Schier, 2019), YouTube (Hupli et al., 2019), online dating sites (Wang, 2020), travel (Terziyska and Damyanova, 2020), social media (Loureiro et al., 2019), brand equity (Chakraborty and Bhat, 2019), video streaming (Wang, 2019), online grocery (Singh, 2019), and consumer engagement (Brodie et al., 2013). Prior studies have neglected to understand consumer participation through nentographic study. The current study tries to bridge this gap.

9.3 RESEARCH METHODOLOGY

9.3.1 Consumer electronic products

This study focuses on consumer electronics because in India, it is the most selling product segment in digital platforms. In general, people are more prone to write and read about consumer electronics because usually companies disseminate new information on electronic products. Therefore, consumers are keen to apprise them before purchasing electronic products. In the domain of various consumer electronics brand communities, people like to share information on consumer electronics which helps the other community members to get updated information and also helps them to make the right purchase decisions.

9.3.2 Brand community of Facebook social media

To select suitable brand communities for conducting netnographic study, 70 respondents were asked about consumer electronics brand community that they prefer the most. According to 70 respondents, the top three Facebook brand communities that they follow were – “Apple Users India”, “Nikon India Photographers”, and “Canon India Photographers”. Later, one researcher joined all the three brand communities. To examine the idealness of three online brand communities, this study compared brand communities with Muniz Jr’s (2001) brand community markers, namely, consciousness of kind, shared rituals and traditions, and moral responsibility. The next step is to analyze whether these three brand communities follow Kozinets’s (2002) guidelines with regards to the suitability of conducting netnography for examining consumer social participation in brand communities.

9.3.3 Entrée Kozinets’ guidelines

Relevant – Here, the communities are based on three brands of consumer electronics. Those are Apple Users India, Nikon India Photographers, and Canon India Photographers. In these three brand communities, members furnish and hunt for information on the consumer electronics. Hence, these communities are relevant.

Active – The deliberations among members are on a high scale. On a daily basis, the members from these communities are engaged in sharing pictures and comments.

Interactive – Community members from these three brand communities are highly interactive. They keep on sharing information on brands in brand communities.

Substantial – There are numerous communicators in these brand communities. In general, several members keep on sharing information in brand communities.

Heterogeneous – From the Facebook profiles of the members, it is observed that they are from various states of India.

Data-rich – Information in these three brand communities is detailed and extensive.

9.3.4 Data collection

It was ensured that the three brand communities were at par with Kozinets' entrée guidelines. The next step was the quotes, reviews, or postings of three brand communities – "Apple Users India", "Nikon India Photographers", and "Cannon India Photographers" – were read and collected manually. The posts collected from brand communities were from Dec 1st, 2023, to Dec 30th, 2023.

9.4 RESULTS

This study conducted three studies differently to understand people's participation in consumer electronic brand community social media that ultimately affects the brands.

9.4.1 Brand community markers

To determine whether "Apple Users India", "Nikon India Photographers", and "Canon India Photographers" are ideal brand communities, this study compared the communities with Muniz Jr's (2001) three brand community markers. The three brand community markers are consciousness of kind, shared rituals and traditions, and moral responsibility.

9.4.1.1 Marker 1 – Consciousness of kind

In general, community members prefer to write "We" or "Us" that ultimately reflects the sense of belongingness. Some of the examples in three brand communities of "Consciousness of kind" are given below:

9.4.1.1.1 Study 1 – Apple Users India

"Hey frnds waned to knw is there any app which is good for ringtone for I phone which we can easily access set as ringtone. (sic)"

"So if we are in india, we cant buy apple accesories from Apple website? (sic)"

9.4.1.1.2 *Study II – Nikon India photographers*

"On the occasion of World Wildlife Day....Our sole interest is to connect you...."

"When we traced them ...put the cameras in the rain covers and had to wait for the weather to get nicer."

9.4.1.1.3 *Study III – Canon India Photographers*

"Your pics are too good. It would be worthwhile to educate all of us thru a YouTube video. I am sure it would be useful. (sic)"

"Sometimes there are things so unique and exquisite that we would miss out on if we didn't stop and focus or zoom in on them!"

9.4.1.2 *Marker II – Shared rituals and traditions*

Similar kind of rituals, traditions, and behavior within the community members remains an important marker. In these three brand communities, the presence of shared ritual and traditions is also observed which are discussed below:

9.4.1.2.1 *Study I – Apple Users India*

Generally, members of the community use hashtags. The hashtags facilitate other community members to understand about topics discussed in respective posts. Another observation is whenever they have queries pertaining to the brand; they mention "thanks in advance" or "TIA" after their queries.

Some of the observations are given below:

> #Mac
>
> Need to record my screen when I'm running Cisco Webex with the usual recording options disabled via that software.
>
> Any third-party option?
>
> Nothing shady - this is for something academic.
>
> (sic)

> Need help with the following scenario:
>
> 1. iPhone set to silent (ringer off)
> 2. Whatsapp notifications on Apple Watch turned ON .... I am missing all my Whatsapp notifications due to this. Can anyone throw some light on how to fix the setup? TIA. (sic)

9.4.1.2.2 Study II – Nikon India photographers

Members generally write about the settings of camera when they click any object. Even members mention the place where the photographs are clicked. Moreover, usually members mention the photographers' names, date of capturing the photographs, and some descriptions about the photographs.

> Portrait
> 30 jan 2020
> iso 400 f /2.8 1/160s+1ev
> 200mm Nikon D500
> Lokhandwala Andheri Mumbai India.
>
> Beauty is how you feel inside,
> and it reflects in your eyes…
> II NIKON D5300 II 70-300 II
> Kolkata 27.03.2020

9.4.1.2.3 Study III – Canon India Photographers

In general, when members share any pictures in the community, they mention the camera settings while clicking those pictures. Some of the observations are given below:

> Desert fox
> Little Rann of Kutch (Gujarat)
> January 2020
> Canon 80d +55-250mm
>
> queen Of Ranthombore Tiger Reserve Quenching Her Thirst. February 2019.
> Canon EOS 5D Mark iv Paired With Sigma Sports Lens 150 600 f 5.6 f 5.6 S S 1/ 500s I S O 5000. F L 150 mm

9.4.1.3 *Marker III – Moral responsibility*

Moral responsibility acts as a measuring rod to assess the level of moral obligations toward each other. Usually, the members believe in assisting each other on several issues on brands. Some of the observations are provided below:

9.4.1.3.1 Study I – Apple Users India

> I have this issue please I was using bootcamp to install windows 10 on a MacBook pro computer, I got an error massage indicates that the hard disk cannot be partitioned…. I found that the system volume is

> unmounted and the options is dimmed The data on the drive is very important and I think their is no backup on the icloud. (sic)

Other members of the community replied to assist that member:

> Maybe you can get Ubuntu on a usb mount the Apple hard drive on it recover your data and do a fresh install of Mac OS (sic)

Another member's response to assist that member:

> Use any other Mac and install the OS on either a 16gb pendrive or hard drive. Connect to this one and boot from it. The internal hard drive becomes external so u can copy the data (sic)

One more observation of a member looking for assistance from community members:

> #Mac

Need to record my screen when I'm running Cisco Webex with the usual recording options disabled via that software. Any third-party option?

One member's response to the query:

> Nothing shady - this is for something academic

Another member's also responded to assist the member:

> Start recording before the app opens?

One more member also assisted:

> You could try screen recording through QuickTime or OBS.

9.4.1.3.2 Study II – Nikon India Photographers

One member is looking for some assistance from community members:

> Issue with my Nikon D5100, while taking picture big shadow is coming where do I rectify this please help.

One member's response to the query:

> Shutter movement issue, i had same issue and changed the shutter (sic)

Another member's response to the query:

> Does this happen when you shoot with flash or without it?
> If with flash then it might be an issue with sync speed. Check your camera sync speed.
> If it happens without flash then shutter itself might have issues and needs repair.

One more member's response to the query:

> Shutter speed 125 kr ke dekh (sic).
> (try setting shutter speed to 125)

From his sense of responsibility, one member shared one video tutorial on photography to assist other members:

> One of the common issues with challenging light situation.... The image has been completely corrected in Adobe Camera.... Though I always advice to get the exposure right in the field, there are workarounds available...

9.4.1.3.3 Study III – Canon India Photographers

One member post for seeking assistance from group members:

> I am planning to buy a camera... Shortlisted 2... Please help which one i should go for... 1500D or 3000D?? (sic)

One member replied for assistance:

> 3000D is good bhai (3000D is good, brother)

One more member replied for assistance:

> Go for full frame camera

Another member is also seeking for assistance:

> Can i use 18 - 200mm lens in Canon M50

One more member replied for help:

> Yes I think you would need adapter...

One more member also replied for assistance:

"M50 ka mount m mount he adapter honese astimal kar saktahe viltrox me sirp full frem ka lens atahe (sic)" (The M50 has a 6.5″ mount and can be fitted with an adapter, the Viltrox only has full frame lenses)

Thus, "Apple Users India", "Nikon India Photographers", and "Canon India Photographers" brand communities have complied to Muniz Jr's (2001) markers. Therefore, these brand communities satisfy to the prescribed guidelines.

The following step is to analyze whether these communities follow Kozinets' (2002) procedures for applying netnography.

9.4.2 Entrée Kozinets' guidelines

1. Relevant – It is observed that the members of "Apple Users India", "Nikon India Photographers", and "Canon India Photographers" brand communities focus on the consumer electronic product category. In these three brand communities, members are giving and seeking opinions on their respective brands. Therefore, it is evident that all these three brand communities are relevant for the study.
2. Active: The activity within these three brand communities "Apple Users India", "Nikon India Photographers", and "Canon India Photographers" are high. On a daily basis in "Apple Users India", numbers of posts in the brand community are from 6 to 8 comments. For "Nikon India Photographers", numbers of posts in the brand community are from 8 to 10 comments. For "Canon India Photographers", numbers of posts in the brand community are from 10 to 12 comments.
3. Interactive: The interactive level in these three brand communities is high. Members are very engaged in exchanging their opinions through various posts.
4. Substantial: There are several active participants in these three communities. The total number of active members in "Apple Users India", "Nikon India Photographers", and "Canon India Photographers" are 7484, 34,850, and 14,896, respectively.
5. Heterogeneous: From the Facebook profiles of these three communities, it is revealed that members belong to different geographical states of India.
6. Data-rich: The posts and observations in these three communities are comprehensive. In general, each post in "Apple Users India", "Nikon India Photographers", and "Canon India Photographers" are around 12 words, 14 words, and 15 words, respectively.

It is evident that all these three brand communities follow Kozinets' (2002) recommendations. Hence, communities are appropriate for applying netnography.

9.4.3 Evidences of consumer's social participation

9.4.3.1 *Informational participation*

Informational participation is defined as the degree to acquire and share new information for the fulfillment of general interest of the brand community (Nadeem et al. 2019).

9.4.3.1.1 Study I – Apple Users India

"Niyo Nylon Band Compatible with Apple Watch Series 5, Series 4, Series 3, Series 2, Series 1"

"#Apple has officially announced the all-new #MacBook Pro supporting a Retina display, 9th generation processor, up to 64GB RAM, up to a whopping 8TB storage....The thermal design on the new MacBook Pro now features a larger impeller...."

9.4.3.1.2 Study II – Nikon India Photographers

"Amazing Thousands Of Ready To Use Instant One Click ***BOKEH LIGHT FILTER*** in Picasso Dg Photo Album Software."

"Are you in the habit of shooting BLURRY IMAGES and don't know how to fix them? You may be using the WRONG SHUTTER SPEED!... Get your basics straight...."

9.4.3.1.3 Study III – Canon India Photographers

"Canon 5DMIII, Canon 100mm f/2.8 macro, f/7.1, 1/200s, ISO 100.Godox AD600E strobe as key light.Godox 500w LED light on godox reflector as shadow fill light. (sic)"

An high key image of the Orangutan and it's baby....This can be achieved by using evaluative metering with an over exposure of +2 stops or spot metering the subject at zero EV (depends on the tonality of the subject you are metering).

9.4.3.2 Actionable participation

Actionable participation can be referred to the degree to which members participate in the community frequently and enhance the level of interactions among themselves in the brand community (Nadeem et al., 2021).

Some of the observations are given below:

9.4.3.2.1 Study I - Apple Users India

One member within the brand community posts Huawei CEO uses Apple product. This post brings a sense of pride which influences community members' engagement with the brand.

> "Huawei CEO, Ren Zhengfei seen with an Apple iPad at Airport security line

He has previously told the press that his children prefer to use Apple products rather than Huawei's own. He also said that he buys them iPhones.

Such an Open Minded person"

One member is looking for someone who needs a screen protector:

"Any iPad Pro users interested in getting Paperlike Screen protector? Just buying one is expensive. If I buy three then shipping is free. (sic)"

Members replied to the above query:
One member replied:
"Check in aliexpress (sic)"

Another member replied:
"are you buying from paperlike.com?"

Study II – Nikon India Photographers
One member's post on photography workshop:

"Hello Friends!
On the occasion of World Wildlife Day ....I request you all to take out a day from your mundane life and attend our Workshop ....All Candidates will be awarded a Certificate from WWF ...."

One member's post on street photography:

" "Hello ShutterBugs, Does street photography excite you?Does idea of photography at festivals of India increase the adrenaline in you?

And you feel nervous to execute your plans about photography, right? ....Trust us... It is not just a Photo Tour. .... photography techniques in the classroom, Guidance on the field, and Analysis of the clicks post the fieldwork. (sic)"

9.4.3.2.2 Study III – Canon India Photographers

A member post on photography activity:
"Hello Photolovers!!!
....We will cover Street and Famous Chandannagar Heritage Buildings and Architecture...."
Another post on Wedding Photography Workshop:
"Canon Image Academy* Advance Wedding Photography Workshop...."

9.4.3.3 Attitudinal participation

Attitudinal participation can be described as a psychological tendency of members to evaluate the brand performance which is shared within the brand community (Kamboj, 2019).

9.4.3.3.1 Study I – Apple Users India

One member wrote his negative emotions toward Apple products:

> ...where most people are going the other way...I moved to the iPhone after only ever using androids....I spend an hour a week cleaning up sms messages, something that just used to happen automatically. Lot of small features are just not present....

Another member accepted the above member's stand. With sarcasm, the member replied:

"Set an alarm, save it...still doesn't say the time left for the alarm to trigger...oh Apple is this really too hard to implement?"

Another member also supported the above member's stand and replied:

"I just bought iPhone 11 as my first iOs device. I regret this now. My 15k android is much much better than this 70k monstrosity (sic)"

One member shared his positive emotions toward Apple products:

> I' m surprised with the battery performance of iPhone 11....I unplugged me phone yesterday morning at 10 AM with 99% battery and today morning still it is at 61%. I am doing the same thing/task which I was performing on my old 6S which I was supposed to charge at-least twice a day. This time Apple you people nailed it. (sic)

There are two members who shared a similar stand on the brand:

"I'm very happy with the battery life on my 11 as well. Easily lasts more than a day and a half!"

"All thanks to IOS 13. battery performance is decent post update."

Study II – Nikon India Photographers

One member shared his exciting experience with Nikon:

> ....An afternoon shot. Light was a challenge but the green background was a plus. Little adjustment of while white balance and green magenta factor, D-Light option in #nikond750 and of course some post-processing helped me get a decent image.

@Sept 2019, Masai Mara. Join #wowxplore Wildlife Expeditions for a knowledge based wildlife experience."

By using a Nikon camera, the member feels he can capture nature perfectly. The post shows his satisfaction with the brand:

"After God perfected the sunrise, he created photographers, artists, and poets to ensure his feat remained immortal.

Terri Guillemets.

Gear - Nikon, Sigma & Benro"

9.4.3.3.2 Study III – Canon India Photographers

One member shares his romantic feelings with his photography:

"Canon 6D EF 50 mm f/1.8 Lens: A single bloom.!!! Love at first sight.

Day end with perfection sunshine sunset behind the beautiful structure of tower

Canon 1500d"

Another member who owns Canon shares nature's glory through his Canon:

"Smell the sea, and feel the sky. Let your soul and spirit fly.With Canon 200d mark ii kit lense 18-55...."

The researchers have used Google translate for the conversion from Hindi to English language.

9.5 DISCUSSIONS OF THE FINDINGS

From the three studies, it is evident that members' informational participation is present. In the above studies, it is observed that some members act like opinion leaders as they constantly provide new information on the brand. From the posts, it is observed that the marketers can recognize the opinion leaders who are known for their pro-activeness and stock of ideas and knowledge. These leaders within the brand community become instrumental in helping the marketers to advertise their products or services. The marketers can motivate these opinion leaders by providing gifts and incentives so that they can influence others on different social media platforms. As they are non-commercially driven, the credibility of the information is very high (Wang et al., 2020). Also, the marketers plan for educational campaigns which ultimately satisfy the customers' informational needs.

From the posts, it is evident that members' actionable participation is present. In the above studies, it is found that some members are insightful in terms of addressing community members' issues. It is observed that some members are skilled in predicting the situation which ultimately proves to be an asset for the community. Even certain posts reveal that some members are articulate enough so that other members can take necessary action plans. All these activities improve the levels of brand engagement of brand community members. The higher the engagement, the higher the acceptance of that particular brand (Stephen et al., 2020). Higher acceptance leads to a rise in customers' motivational spirit. Moreover, their behavior in terms of cognitive and emotional affairs is also influenced.

From the above-mentioned posts, it is clear that the members' attitudinal participation is present. Members exhibit their sense of happiness by showcasing the brand. If members' opinion is positive, it will foster a sense of acceptance among other members. On the contrary, if the opinion is negative, the marketers can come forward for understanding and correcting the products or service as far as the need and demand from the members are concerned.

9.6 CONCLUSIONS

The study examined customer social involvement in Facebook brand communities. UGT was used to evaluate three forms of consumer social participation. Sharing consumer electronics facts or opinions pleases them. Consumers feel obligated to counsel or lead other community members by providing favorable or bad opinions or ambivalence. Members share unfavorable messages to guide other community members and marketers on brand problems.

Netnography study focuses on consumer electronics because it's the most popular product area, and there are many online opinions about it. Netnography surveys show customer mood for any brand. Three consumer electronics brand communities were selected by 70 Indian respondents. They were Apple Users India, Nikon India Photographers, and Canon India Photographers. Comparing the brand communities to Muniz Jr's (2001) principles helped determine if they were ideal. This ensured that all three brand communities have Muniz Jr (2001) identifiers. Ideal brand communities have three brand community identifiers. The next stage was to assess community appropriateness for netnography. Compare brand communities to Kozinets' Entrée rules. After establishing brand communities suited for netnography, the study examined customer social involvement (informational, actionable, and attitudinal). The study revealed all three customer social involvement elements in brand community postings.

9.7 THEORETICAL IMPLICATIONS

This study unearths the psychological drivers of consumer's behavioral patterns in online social media. The results of the study claim that UGT facilitates the sharing and accessibility of various information through online social platforms. People accumulate motivation and social capital by exchanging various forms of information. The study also reflects the consumer's motivational factors behind UGT used as the theoretical backbone of the study. Prior studies mainly followed UGT to understand the gratification by using certain social media platforms or applications. However, the extant literature fails to extend UGT in terms of people's behavioral participation. Through this study by extending UGT or in other words, gratifications of consumer participation which ultimately facilitates the brand to identify the opinion leaders to improve brand engagement level of consumers and the performance of the brand. This study uncovers the gratifications behind consumer social participation in online brand communities.

9.8 MANAGERIAL IMPLICATIONS

From the informational posts, marketers can identify the opinion leaders of the brand. They are the early users of the brand as well. The marketers can influence them by providing incentives so that they can influence others to buy the marketers' brands on various virtual platforms. Their opinions are considered credible because of their immense acceptance and non-commercial traits. The marketers can conduct various educational campaigns on the brand in regular intervals. It will enhance the consumer engagements in online brand communities. They can establish the environment of trust, empowerment, consumer value, and satisfaction in the virtual brand communities. It can bring a transient consumer engagement which reflects a broader, dynamic, social interaction, bonding, and interactive engagement process. Since consumer engagement is multidimensional – cognitive, emotional, and identification (Hollebeek, Srivastava, and Chen, 2019), marketers in the community can be more attentive to address these compositions of engagement, keeping the long-term relationship in mind which further is required for the success of business. Efforts can be taken to disseminate the brand story and image to its customers with a system of interactive feedback for further improvements.

In order to enhance social participation in the virtual brand community, marketers can use various posts to enhance the brand positioning. They can give a free hand to community members to write various posts freely on the brand. When members share positive opinions, marketers can track those people and influence them so that they can talk positively about the brand on other social media platforms as they are non-commercially driven. The credibility of information that they are sharing will be well accepted by other people. On the contrary, when members share negative opinions about brands, the marketers can exactly understand the error in the brand. By consulting with R&D, the marketers can rectify those errors which can change the negative attitude of people to a positive one.

9.9 LIMITATIONS AND FUTURE RESEARCH DIRECTIONS

Indian consumer electronics is the topic of this investigation. Social media platforms may be used for global studies. We can duplicate the study in additional product or service categories. Cross-culture studies can examine brand community social involvement by considering different products or services. This study uses qualitative methods. Studies using quantitative or hybrid methods can increase generalizability and add to current information. This analysis considered commercial brands. To extend study horizons, investigate modern and creative social goods or causes. Medium study on YouTube, Twitter, and Telegram may assess the influence of social engagement and compare the best medium for informal business communication.

This UGT-based study examines brand community Facebook engagement. Since brand communities are overwhelmed with information through postings, consumers feel dissonant or anxious about what to utilize and what to dismiss. To assess involvement, Leon Festinger's, 1957 cognitive dissonance theory might be applied. Such theoretical contributions will advance management, economics, communication, and psychology research.

Some social media ideas have particular places in information systems. According to social information processing theory, online relationships can be as successful as offline ones. Growing relationships might be stronger than offline ones. Another argument is that social media helps communities build social capital, which enhances economic capital over time. Another hypothesis, shared-world theory, states that social media posts build communities. This research examined brand community social involvement without exploring media and communication studies.

REFERENCES

Abid, Aman, and Paul Harrigan. "An exploration of social media-enabled voter relationships through uses and gratifications theory, psychological contract and service-dominant orientation." *Australasian marketing journal* 28, no. 2 (2020): 71–82.

Brodie, Roderick J., Ana Ilic, Biljana Juric, and Linda Hollebeek. "Consumer engagement in a virtual brand community: An exploratory analysis." *Journal of business research* 66, no. 1 (2013): 105–114.

Chakraborty, Uttam, and Savita Bhat. "Are online opinion leaders and seekers distinct? A study on consumer electronics industry in India." *Global Business Review* 20, no. 3 (2019): 813–825.

Creswell, John W., and Cheryl N. Poth. *Qualitative inquiry and research design: Choosing among five approaches*. Sage publications, 2016.

Dhir, Amandeep, Gina M. Chen, and Sufen Chen. "Why do we tag photographs on Facebook? Proposing a new gratifications scale." *New Media & Society* 19, no. 4 (2017): 502–521.

Eaton, Paul William, and Laura A. Pasquini. "Networked practices in higher education: A netnography of the# AcAdv chat community." *The Internet and Higher Education* 45 (2020): 100723.

Festinger, L. 1957. *A theory of cognitive dissonance*, Stanford University Press.

Florenthal, Bela. "Young consumers' motivational drivers of brand engagement behavior on social media sites: A synthesized U&G and TAM framework." *Journal of Research in Interactive Marketing* 13, no. 3 (2019): 351–391.

Hollebeek, Linda D., Rajendra K. Srivastava, and Tom Chen. "SD logic–informed customer engagement: integrative framework, revised fundamental propositions, and application to CRM." *Journal of the Academy of Marketing Science* 47 (2019): 161–185.

Huang, Jialiang, and Liyun Zhou. "Timing of web personalization in mobile shopping: A perspective from Uses and Gratifications Theory." *Computers in Human Behavior* 88 (2018): 103–113.

Hupli, Aleksi, Moritz Berning, Ahnjili Zhuparris, and James Fadiman. "Descriptive assemblage of psychedelic microdosing: Netnographic study of Youtube™ videos and on-going research projects." *Performance Enhancement & Health* 6, no. 3–4 (2019): 129–138.

IBEF. 2024. E-commerce Industry in India: The Indian e-commerce market is expected to have a gross merchandise value of US$ 350 billion by 2030. https://www.ibef.org/industry/ecommerce (accessed May 23, 2024).

Statista. 2024. Leading countries based on Facebook audience size as of April 2024. https://www.statista.com/statistics/268136/top-15-countries-based-on-number-of-facebook-users/ (accessed May 23, 2024).

Islam, AKM Najmul, Matti Mäntymäki, and Hajer Kefi. "Decomposing social networking site regret: a uses and gratifications approach." *Information Technology & People* 33, no. 1 (2020): 83–105.

Jayasingh, S. "Consumer brand engagement in social networking sites and its effect on brand loyalty". *Cogent Business & Management*, 6 (1), 1–22. (2019).

Kamboj, Shampy. "Applying uses and gratifications theory to understand customer participation in social media brand communities: Perspective of media technology." *Asia Pacific Journal of Marketing and Logistics* 32, no. 1 (2020): 205–231.

Katz, Elihu, Jay G. Blumler, and Michael Gurevitch. "Uses and gratifications research." *The public opinion quarterly* 37, no. 4 (1973): 509–523.

Kaur, Harsandaldeep, Mandakini Paruthi, Jamidul Islam, and Linda D. Hollebeek. "The role of brand community identification and reward on consumer brand engagement and brand loyalty in virtual brand communities." *Telematics and Informatics* 46 (2020): 101321.

Kim, Bumsoo, and Yonghwan Kim. "Facebook versus Instagram: How perceived gratifications and technological attributes are related to the change in social media usage." *The social science journal* 56, no. 2 (2019): 156–167.

Kim, Myung Ja, Choong-Ki Lee, and Noshir S. Contractor. "Seniors' usage of mobile social network sites: Applying theories of innovation diffusion and uses and gratifications." *Computers in Human Behavior* 90 (2019): 60–73.

Kozinets, Robert V. "On netnography: Initial reflections on consumer research investigations of cyberculture." *Advances in consumer research* 25, no. 1 (1998): 366–371.

Kozinets, R. V. 2010. *Netnography: Doing etnographic research online.* Thousand Oaks,CA: SAGE.

Kozinets, Robert V. "The field behind the screen: Using netnography for marketing research in online communities." *Journal of marketing research* 39, no. 1 (2002): 61–72.

Kozinets, Robert V., and Ulrike Gretzel. "Netnography evolved: New contexts, scope, procedures and sensibilities." *Annals of Tourism Research* 104 (2024): 103693.

Kepios. 2024. Digital 2024: India. https://datareportal.com/reports/digital–2024-india (accessed May 23, 2024)

Lim, Heejin, and Archana Kumar. "Variations in consumers' use of brand online social networking: A uses and gratifications approach." *Journal of Retailing and Consumer Services* 51 (2019): 450–457.

Linsenmeyer, W., and H. Schier. "Nutrition-related messages shared among the online transgender community: A netnography of YouTube vloggers." *Journal of the Academy of Nutrition and Dietetics* 119, no. 9 (2019): A70.

Liu, Xiaodan, Qingfei Min, and Shengnan Han. "Understanding users' continuous content contribution behaviours on microblogs: An integrated perspective of uses and gratification theory and social influence theory." *Behaviour & Information Technology* 39, no. 5 (2020): 525–543.

Loureiro, Sandra Maria Correia, Jessica Serra, and João Guerreiro. "How fashion brands engage on social media: A netnography approach." *Journal of Promotion Management* 25, no. 3 (2019): 367–378.

Luo, Nuan, Mingli Zhang, and Wenhua Liu. "The effects of value co-creation practices on building harmonious brand community and achieving brand loyalty on social media in China." *Computers in Human Behavior* 48 (2015): 492–499.

Muniz Jr, Albert M., and s. "Brand community." *Journal of consumer research* 27, no. 4 (2001): 412–432.

Nadeem, Waqar, Mari Juntunen, Nick Hajli, and Mina Tajvidi. "The role of ethical perceptions in consumers' participation and value co-creation on sharing economy platforms." *Journal of Business Ethics* 169 (2021): 421–441.

Ray, Arghya, Amandeep Dhir, Pradip Kumar Bala, and Puneet Kaur. "Why do people use food delivery apps (FDA)? A uses and gratification theory perspective." *Journal of retailing and consumer services* 51 (2019): 221–230.

Singh, Reema. "Why do online grocery shoppers switch or stay? An exploratory analysis of consumers' response to online grocery shopping experience." *International Journal of Retail & Distribution Management* 47, no. 12 (2019): 1300–1317.

Labrecque, Lauren I., Kunal Swani, and Andrew T. Stephen. "The impact of pronoun choices on consumer engagement actions: Exploring top global brands' social media communications." *Psychology & Marketing* 37, no. 6 (2020): 796–814.

Tavakoli, Rokhshad, and Sarah NR Wijesinghe. "The evolution of the web and netnography in tourism: A systematic review." *Tourism management perspectives* 29 (2019): 48–55.

te Lintelo, Dolf JH, Kevin Hernandez, and Rajith Lakshman. "Migrant digital placemaking for wellbeing: A netnography of Facebook groups." *Wellbeing, Space and Society* 6 (2024): 100196.

Terziyska, Ilinka, and Radina Damyanova. "Winescape through the lens of organized travel–a netnography study." *International Journal of Wine Business Research* 32, no. 4 (2020): 477–492.

Wang, Yi-Sheng. "User experiences in live video streaming: a netnography analysis." *Internet Research* 29, no. 4 (2019): 638–658.

Wang, Yi-Sheng. "The application of netnography to the online dating service experiences of female users." *Behaviour & Information Technology* 39, no. 2 (2020): 226–240.

Wang, Zongshui, Haiyan Liu, Wei Liu, and Shouyang Wang. "Understanding the power of opinion leaders' influence on the diffusion process of popular mobile games: Travel Frog on Sina Weibo." *Computers in Human Behavior* 109 (2020): 106354.

Chapter 10

Data-driven decision-making approach on healthcare services

A post-pandemic Indian perspective

Hemant Gupta, Pankaj Dixit and Bhaveshkumar J. Parmar

10.1 INTRODUCTION: BACKGROUND AND DRIVING FORCES

The widespread global health concern among people in all countries is the novel corona virus (Covid-19) which was announced to be a pandemic on March 11, 2020, by the World Health Organization. In the latter half of 2019, a new acute respiratory illness known as Covid-19 was discovered. It features symptoms similar to the flu, such as a high temperature, cough, trouble breathing, exhaustion, and sore muscles. It has manipulated the social lives, economic lives, political lives, and cultural lives of people all over the world. The revelation that India's healthcare system is inadequate due to the spread of the Covid-19 virus has brought to light the critical need for fundamental systemic changes (Singh & Parida, 2022).

Since the coronavirus first appeared in 2020, it has been responsible for a level of disruption in human life that has never been seen before. Because Covid-19 has brought to light the inadequacies in India's healthcare system, it has caused difficulties for patients as well as for the medical personnel and facilities that provide treatment. Most nations with few resources have healthcare systems that are ill-equipped to deal with the problems posed by the Covid-19 pandemic. These countries' healthcare systems suffer from significant funding constraints, low equity, and poor quality (Yiu *et al.*, 2021).

Keeping in view the pervasive equivocation, community stress, remain-at-home instructions, and other restrictions imposed during the early phase of the global pandemic in April 2020, it was hypothesised that several patients suffered from acute medical disorders. India's healthcare system has become increasingly centralised, costly, and impersonal over time. Such treatment is unreachable, unresponsive, and costly in a country where healthcare expenditure is low, the majority of healthcare expenditure is out-of-pocket, and the majority of the population continues to reside in rural regions or on the urban periphery. The Covid outbreak further highlighted these truths. In addition, it has made it imperative to rebuild the system to be more decentralised, dispersed, and responsive. During the outbreak, healthcare systems across India were significantly overburdened. Unprecedented situations,

DOI: 10.1201/9781003621102-10

such as the present Covid pandemic, exert immense pressure on healthcare organisations to reorganise hospital facilities and regulations to avoid spreading deadly diseases and ensure effective treatment administration (Estrela *et al.*, 2023).

Positive and negative feelings were experienced by Covid-19 inpatients. They attributed the unfavourable experiences to organisational obstacles and medical staff shortages, particularly among clinical nurses. Patients' expectations are reasonable and proportional to their circumstances. In addition to professional treatment, people demand accurate advice about what to do following hospital release. A deeper comprehension of the patient experience may enable clinical nurses to give better care. Nevertheless, these obstacles have intensified during the first and second rounds of Covid hospitalisation. Patients' economic status and corresponding preconceptions from medical centres, coupled with the quality of healthcare services provided, are substantial patient-specific variables in patient satisfaction; it is established that the level of service quality tends to decrease with a longer in-patient stay (Buzgan and Güner, 2020).

The financial implications of hospitalised patients' care during a pandemic may be a significant element in their reported degree of satisfaction, given that quality increases need higher expenses (or, conversely, cost reductions may result in worse quality). As a result, the patient's perceived financial analysis may, therefore, have an important impact on the patient's experience with medical treatment. The expense of treatment and patient experience might have a weak association, especially during the economic downturn caused by the pandemic. In the context of Covid-19, where financial difficulties exist at practically every level, moderate cost or complimentary treatment can be a more significant component in enhancing patient satisfaction. As a result of Covid, the healthcare industry has experienced significant shifts. It enables us to improve parts of healthcare in a post-Covid era. Without doubt, it will alter the patient's impression of private institutions. It raised the question of whether the type of facility is best suitable for the variety of treatment patients may receive in hospitals. This issue was brought to the attention of patients, health service providers, insurance providers, governments, and other authorities by Covid (Wu *et al.*, 2022).

This chapter summarises the results of an empirical study done to establish the elements that Indian patients deemed significant while selecting a hospital. This study is relevant and significant because Indian consumers have become more aware and affluent in recent years, and their preferences and tastes, as well as their product and service preferences, have evolved through time. As the Covid-19 epidemic continues, understanding patients' opinions of hospital care will play a crucial part in decision-making and redesigning hospital services to boost patient care. To maintain and improve the standard of medical and spiritual care, it is crucial to listen to patients and collect and analyse data on their experiences. To provide safe treatment during pandemics, medical personnel require new capabilities. Rare qualitative research cannot be generalised to the experiences of Covid-19 hospitalised patients. Due to the paucity of research on the perspectives of hospitalised patients, we attempted to address this deficiency by examining the viewpoints of patients. The purpose of this study was to determine the problems and expectations of Covid-19 hospitalised patients.

Accordingly, this study is guided by the following research questions:

- What selection characteristics do customers consider essential when selecting hospitals during the Covid-19 pandemic?
- How do different aspects of a service stack up against one another as selection criteria?
- How do customer demographic characteristics relate to selection criteria and service?
- When patients use hospital services, what kind of reactions do they have?

10.2 LITERATURE REVIEW

Connection to a residential area was noticed to be the most influential concern, preceded by advanced technologies (structure, hygiene, lifts connect, environment), brand and image, expense, healthcare, administrative documentation, and accessibility of medical specialists and assistance. Whitehead *et al.* (2007) did extensive research at a National Health Service (NHS) facility to explore which variables influenced patients' perceptions of sanitation. Thus, according to their observations, the hygiene of the facility is a significant component that patients consider when selecting an NHS hospital. The research arrived at the following conclusions. Consumer evaluations of hospital prestige and health services have a significant role in hospital selection. Further, people preferred to choose hospitals with significant clinical quality monitoring.

Furthermore, contentment with a previous stay in the hospital has a significant influence on later hospital selection. As per the report, a couple of miles from the hospitals was generally the most important consideration in hospital selection. For our study, two groups of earlier studies are significant. First, as outlined by Glied (2000), is an examination of governance and cost controls. However, these articles do not investigate the association between hospital pricing and referrals. There are a couple of further recent studies that address comparable concerns. Cutler, McClellan, and Newhouse (2000), for instance, examine the management of cardiovascular disease in health care with regular insurance plans and find that health centres incur 30% to 40% fewer costs. Almost all of the difference is attributable to decreased unit pricing as opposed to variations in actual treatments. Nevertheless, researchers do not analyse whether cost reductions are the result of lower negotiated hospital pricing or patient referrals to less expensive institutions. Gaynor *et al.* (2004) examine how HMOs (Health Maintenance Organization) accomplish cost reductions in further depth. Within a single HMO, they examine physician reactions to cohort financial incentive contracts. The larger the physician group getting group-based incentives, the greater the expenditures on healthcare, especially for outpatient care. In other words, spending is adversely associated with the effectiveness of strategies to limit spending.

It has been found that the far more essential elements in the hospital screening process were medical competence, hospital credibility, physician judgement, and convenience. Singh and Shah (2011) carried out a survey of 143 people from three privately run hospitals in Manipur, India, and determined 11 causes, which include general practitioner and surgeon accessibility, infrastructural facilities, assurance, recommendation by physicians, recommendation from friends, and relatives, mass transit convenience, adjacency to home, expense, media manipulation, customer priority, and insurance coverage facility. The selection considerations stated in other nations may not be as applicable in the Indian context due to cultural differences, climate, national characteristics, insurance plans, and wellness delivery systems. Additionally, analyses carried out in a different scenario may culminate in the revelation of completely undiscovered components.

Since 2020, the proliferation of coronavirus has caused enormous disruptions in human life. Covid-19 has shown flaws in the Indian healthcare system, causing issues for both patients and healthcare professionals/hospitals. It is now time for India to plan for the post-Covid period and consider how best to translate this challenge into a chance to improve our medical system to meet future needs. This may be an opportunity to build India's healthcare service systems not only according to international standards but also as a massive macroeconomic benefactor to the nation. A further concern of the medical community is that we don't abandon non-Covid patients, such as those with chronic conditions such as diabetes, cardiovascular disease, and cancer, which continue to have high death rates despite the presence of Covid and make a significant contribution to mortality in Covid-19 patients suffering from any co-morbidities.

Flexner *et al.* (1981) performed research to discover the elements that influence patients' hospital selection decisions. Existing research on hospital choices is insufficient in identifying the ultimate decision-makers in hospital selection and the criteria on which they choose hospitals. According to the study, before choosing a hospital, customers of medical services typically shop about and make comparisons. Before choosing a hospital, it has been noted that patients consider the service quality, demeanour and conduct of hospital employees, the hospital's reputation, and the facility's cleanliness. Thus, it has been determined that hospital service qualities have a substantial role in hospital selection decisions.

Thereafter, Boscarino and Stelber (1982) developed a set of characteristics that influence patients' hospital selection decisions in an organised manner. These considerations include proximity to residence, accessibility of specialised physicians, cutting-edge technology, a clean and sanitary atmosphere, acquaintance with medical staff, expertise with healthcare, cost of 23 treatments, number of beds, and religious affiliation. They discovered that consumers prefer to make decisions based on these factors as opposed to doctors' advice. In addition, they have enlarged the hospital selection criterion to include the variety and type of services offered by hospitals. They have discovered that patients pick hospitals based on the necessary services. They have divided the service needs of healthcare customers into three categories: *general services, specialised services*, and *emergency services*.

In addition, Barnartt (1990) have theoretically and experimentally investigated the characteristics that individuals deem crucial when selecting a hospital. The research has been conducted in two stages. Patients' decisions on which hospital to go to are affected by several issues, which were explored in the first phase, and then respondents' perspectives on hospitals were addressed in the second. Prior service utilisation, staff friendliness, facility variety, service cost, proximity to home, service standards, service cost, service quality ratings, and physician and peer references were all shown to be important in the selection of a hospital. The authors have summarised these considerations by narrowing them down to four primary characteristics: expertise, price, excellence, and suggestions. The research was carried out on patients from privately owned and two government hospitals in the city of Riyadh. Using a validated questionnaire, primary information was gathered from 540 patients. Utilising exploratory factor analysis, we have identified characteristics that influence patients' hospital selection preferences. To investigate the factors influencing the choice behaviour of both government and private hospital patients, a discriminant analysis with steps was conducted. Patients' hospital selections are influenced by five factors: regulatory processes, the effectiveness of healthcare, hospital reputation, the cost of treatment, and medical insurance. Eight characteristics have been identified as statistically meaningful for distinguishing between governmental and private hospital patients. These factors are therapeutic expense, occupation, quality of service, administration convenience, psychological well-being, medical insurance, age, and sexuality. On the premise of these data, it was determined that facility selection decisions are impacted by institutional, demographic, and socioeconomic characteristics.

Various research studies have indicated that customers of medical care services participate in the selection of a hospital. This research has revealed that patients examine several aspects while picking a facility. For instance, these characteristics include quality of service, specialty physicians, high-tech technology, hospital prestige, doctor and family referrals, and proximity to residence. The majority of the research cited has highlighted hospital selection considerations from the standpoint of a developed nation. Consequently, a research question has been discovered in the current body of literature. These studies have given a foundation for examining the selection factors that customers in a growing nation like India may deem essential when selecting a hospital.

Multiple variables contribute to the patient's satisfaction with healthcare facilities, as evidenced by the findings of the research. Important factors include staff attitude, timeliness of services, staff friendliness and civility, quality of nursing care, and physical infrastructure. The elements that contribute to patient satisfaction in industrialised economies. Consequently, a study vacuum has been identified in the literature regarding patient satisfaction with hospital treatments in prosperous economies (Mohammad Mosadeghrad, 2014).

Findings suggest that India's public healthcare industry is expanding at a pace of 15% each year. There are several causes for such rapid expansion. First, according to the 2011 census conducted by the Government of India, the population has surpassed 1.21 billion, resulting in an increasing need for medical services. Second, medical tourism has significantly increased. India has emerged as one of the most popular locations for overseas patients seeking best-in-class hospital services at a reasonable cost. Third, because of increased income levels, the standard of living of many Indian middle-class families has risen over time, necessitating greater medical treatment. Rapid growth has prompted the establishment of several new private hospitals around the region. Fourth, the rise in lifestyle diseases such as coronary heart disease, diabetes, and obesity has compelled consumers to invest more in medical services. As a result of these primary factors, there has been an expansion of facilities for healthcare in India. Consequently, Indian clients have several options for reviewing healthcare facilities before making their ultimate selection (Mohammad Mosadeghrad, 2014; Ghosh, 2015).

Primary care is the cornerstone of a nation's medical services. The quality of medical care provided to the majority of the people is the best measure of a country's degree of socioeconomic development and establishes the general health of the nation. Patient happiness has been associated with higher patient adherence, continuity of treatment, improved clinical outcomes, and increased service usage and risk management. Thus, patient satisfaction is a crucial measure of the excellence of healthcare provision and an essential metric for evaluating and enhancing healthcare services. In the 1950s, the United States was the birthplace of healthcare quality research. The initial research sought to predict patient happiness by identifying patient variables such as age, gender, and ethnicity (Verma *et al.*, 2020).

10.3 RESEARCH GAPS

Since patients are the consumers of the services offered by healthcare organisations, service providers must have a deeper understanding of their behaviour to utilise the limited resources efficiently. It may aid organisations in designing, producing, delivering, and enhancing services in line with the requirements and preferences of various patient groups. Most investigations have been undertaken in developed nations. The delivery of healthcare systems in these nations varies from the cash-based payment approach prevalent in emerging and poor nations. Regarding culture, patient expectations, the role of the government in medical care, and the demographic features of the respondents, there are several differences across developing nations.

Most of the prior research has explored the behaviour of healthcare customers from a broad viewpoint. They have not focused primarily on patients suffering from Covid-19 at multispecialty facilities. In addition, there is a dearth of research that examines the behaviour of healthcare customers regarding hospital selection, satisfaction levels, post-purchase intents, and decision-makers engaged in hospital selection during the Covid-19 era. This gap in the literature has been identified in both developed and developing nations.

In the context of tertiary-level healthcare services in multispecialty hospitals in India, an emerging country, the goal of this study is to investigate the factors influencing hospital choice decisions and patient satisfaction in the post-Covid era. It evaluates if there is a significant difference statistically between these parameters and the existence or lack of health insurance, as well as between various demographic groups of respondents. It characterises the decision-makers engaged in the choice of hospitals and the recommendation behaviours of patients. It also presents a framework for customer preferences and healthcare industry satisfaction. Therefore, the present work seeks to fill the highlighted gaps and provide theoretical contributions to the existing literature.

10.4 RESEARCH METHODOLOGY

Every research must be well organised. To conduct research methodically and systematically, it is necessary to create an appropriate research technique. This describes the general study plan and the technique used to conduct the research. The present study was undertaken in the context of the Indian post-Covid healthcare system to identify significant characteristics and their relative relevance, which influence patients' hospital choice decisions and their levels of satisfaction with tertiary-level medical services. In addition to analysing patients' readmission and referral behaviours, the study classifies judgements engaged in hospital selection.

The primary facts of the patient will be analysed, with the reasons for selecting the institutions emphasised and organised into logical and understandable categories. Building linkages between and among categories, elucidating the relationship between categories, and developing a theory based on the ties between categories. We will use a mixed methodology (qualitative and quantitative analysis) by using relevant software.

The objective of the study is to determine the most important elements that influence patients' decisions when selecting a healthcare provider in post-Covid scenario; whether there is a correlation between demographic variables and the factors that influence patients' decisions on which medical facilities are related to Covid-19 treatment. This study also aims to isolate and quantify the most influential elements influencing patients' overall satisfaction with healthcare providers, to determine if there is a correlation between demographics and the elements that influence patients' satisfaction with healthcare facilities and to examine the factors that influence patients' decisions to use and suggest certain healthcare providers' services.

10.5 RESEARCH HYPOTHESIS

H_1: Patient selections in healthcare facilities are significantly correlated with various facilities, Covid-19 prevention protocols, and amenities provided by the hospital.
H_2: Patients' overall levels of contentment with healthcare services are significantly correlated with several demographic factors.
H_3: Consumer preferences heavily influence the decision to recommend healthcare services already purchased.
H_4: A high level of customer satisfaction is directly correlated with patients' propensity to refer their friends and family to a healthcare provider.

The research tool was created following an exhaustive and methodical literature evaluation. To facilitate the collection of primary data from respondents, a standardised questionnaire was developed. The questions have been generated methodically. The researcher has first compiled a list of items they need from participants. A set of questions was determined to gather data based on the aims of the current study. The questionnaire has been broken down into five components based on a combination of research and input from a panel of experts. In the first part of the report, we list the names of patients, the relationship of visitors with patients, and the mode of referral to the particular hospital. In Section 10.3, response group demographics are presented. Section 10.4 is a list of factors that influence a patient's decision on which hospital to visit. In Section 10.5, we'll go through the various factors that might affect a patient's level of contentment with their hospital stay. Section 10.6 focuses on the respondents' subsequent behaviours after using the service (see Table 10.1).

Table 10.1 Each question and its intended response technique are listed below

Section no.	*Gist of questions*	*Response method*	*Purpose*
1	Patient name Relation of visitor with the patient Mode of referral to hospital	Open-ended responses	To make identification of patients and decision-makers

(Continued)

Table 10.1 (Continued)

Section no.	*Gist of questions*	*Response method*	*Purpose*
2	Age Sex Residence Occupation Education Distance from home to hospital Family income Health insurance	Please tick one option out of the given different options	To collect the demographic information to study respondents
3	List of attributes related to patient's choice of hospitals	Multiple choice questions to be answered	To identify the factors affecting hospital choice decisions of patients
4	List of attributes related to patients satisfaction from the hospital services	Multiple choice questions to be answered	To identify the factors affecting patient satisfaction with hospital services
5	Satisfaction level Recommendation level	To be chosen from given options as per the experience	To study the patient's decisions for repurchase and recommendation of hospital services

Source: Authors' calculation.

Table 10.2 Demographic criteria and health insurance-wise break-up of respondents

	Gender		*Age*			
	Female	*Male*	*18–45*	*45–60*	*60+*	*Below 18*
Frequency	93	114	106	48	26	27
Percent	44.9	55.1	51.2	23.2	12.6	13
Valid percent	44.9	55.1	51.2	23.2	12.6	13

Source: Authors' calculation.

10.6 DATA ANALYSIS

The distribution of respondents on the basis of demographic categories and health insurance has been shown below (see Table 10.2):

Number of respondents on the basis of demographic criteria and health insurance are mentioned in this section.

- Age – 27 (below 18), 106 (18–45 years), 48 (45–60 years), 26 (more than 60 years)

- Gender – 114 (male), 93 (female)
- Residential rea – 89 (rural), 53 (semi-urban), 65 (urban)
- Education – 38 (not educated), 67 (up to 12th), 67 (graduate), 30 (postgraduate)
- Occupation – 25 (farmer), 32 (govt. employee), 46 (not employed), 48 (private employee), 26 (self-employed), 30 (student)
- Annual family income – 25 (below 2 lakhs), 54 (2–5 lakhs), 90 (5–10 lakhs), 38 (above 10 lakhs)
- Health insurance – 51 (yes), 156 (no)

Out of 207 respondents, 27 (13%) are below 18 years of age, 106 (51%) are in the age group of 18–45 years, 48 (23%) are between 45 and 60 years, and 26 (13%) are above 60 years of age. The numbers of male and female respondents are 114 (55%) and 93 (45%), respectively (see Table 10.3).

As a result of the Covid outbreak in India, this study was designed to investigate how the pandemic has affected consumer preferences and levels of contentment. It's based on a study of people receiving care in multispecialty hospitals in north India, with a focus on those with tertiary-level illnesses. A total of 207 patients were surveyed directly through a questionnaire to provide the primary data used in this study (see Table 10.4).

Table 10.3 Residence- and education-wise distribution of respondents

	Residence			*Education*			
	Rural	*Semi-urban*	*Urban*	*Not educated*	*Up to 12th*	*Graduate*	*Postgraduate*
Frequency	89	53	65	38	67	72	30
Percent	43.2	25.2	31.5	18.4	32.4	34.8	14.5
Valid percent	43.2	25.2	31.5	18.4	32.4	34.8	14.5

Source: Authors' calculation.

Table 10.4 Mode of referral to hospital-wise distribution of respondents

	Mode of referral to hospital				
	Friends	*Relatives*	*Self*	*Social media*	*Newspaper*
Frequency	43	57	64	22	21
Percent	20.7	27.5	30.9	10.6	10.1
Valid percent	20.7	27.5	30.9	10.6	10.1

Source: Authors' calculation.

Healthcare customers' varied attitudes and actions have been studied by considering several demographic factors. The data analysis concluded that there are several factors which contribute to positive patient satisfaction and highlight the expectations of the people from the hospitals, especially after the first and second wave of Covid-19 which uncovered the loopholes in the Indian healthcare system (see Figure 10.1); similarly, it also allowed the healthcare sector to cover the faulty lines and proceeded to the scope of progress (see Figure 10.2).

The frequency distribution of demographic variables explained that education, occupation, annual family income, and health insurance are important factors that contribute to patient choices. Almost all of those who took part in the survey said that basic amenities like a water facility, electricity backup, clean restrooms, emergency exit, lifts, parking facility, housekeeping, an easily accessible medical store, a cafeteria, and a large enough waiting space were all critical in deciding which hospital to visit (see Table 10.5).

Patients' decisions are heavily influenced by the hospitals' brand names, clinicians' reputations, and the hospitals' public image. The selection of hospitals has also taken into account patients' religious and cultural preferences, the hospital's level of experience, and the availability of high-calibre specialised physicians.

10.7 RESULTS AND DISCUSSION

It has been found that patients prefer hospitals which have the availability of cashless payment methods/digital payment systems, adequate laboratory facilities, availability of oxygen and ventilators, in-house diagnostic equipment, and provision of medicines at discounted rates. Cost of services, ease of approaching the hospital, number of rooms, ease of getting appointments,

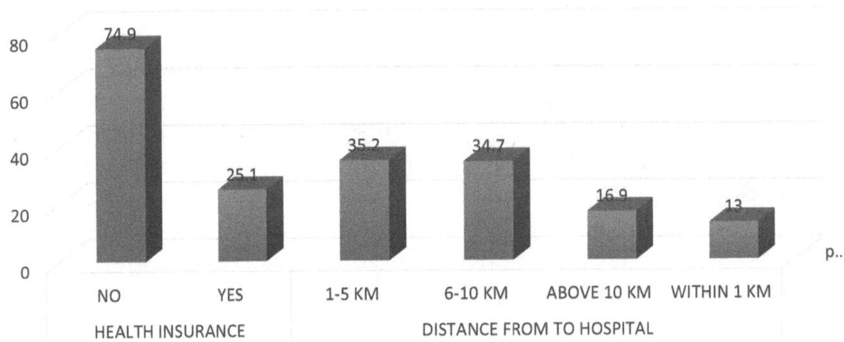

Figure 10.1 Health insurance and distance from home to hospital-wise distribution of respondents.

Source: Authors' calculation

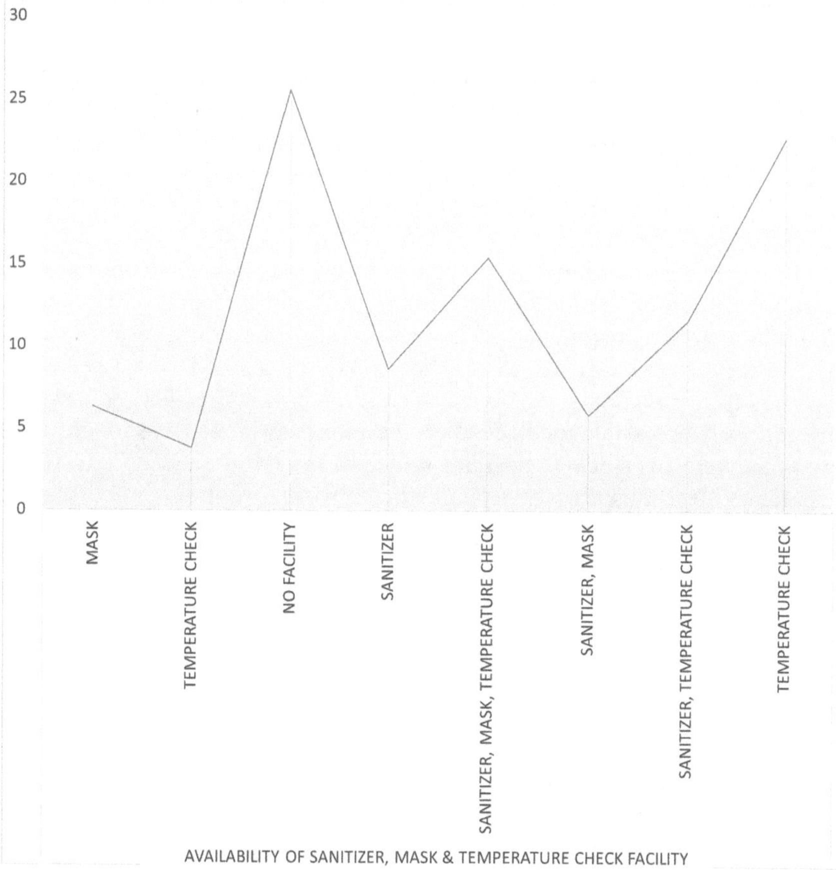

Figure 10.2 Availability of sanitiser, mask, and temperature check facility.

Source: Authors' calculation

hospital's tie-up with insurance companies, nearness to home and number of beds in the hospital also have a significant impact on a patient's final decision. It has been observed that patients generally prefer such hospitals where any friend/relative/known person is working. Timing of OPD (Out Patient Department) services, awareness about disease and treatment procedures, severity and duration of the illness, and previous experience with the hospital also influence patients' decisions. It was observed that patients were quite satisfied with the hospitals having full arrangements for tackling Covid patients such as separate corridors/wards for Covid patients, sanitiser, masks, and temperature check facilities at the entrance of the hospitals. Covid-19 screening facility in the hospital, use of surgical mask/cloth mask or N-95 mask by the physicians and paramedical staff, and use of gloves and sanitisers before examination of the patients.

Table 10.5 Other factors in the distribution of responses

	Registration time			*Mode of payment*			
	15–30 Mins	*>30 mins*	*Upt o 15 mins*	*Card payment*	*Cash*	*Pay later*	*UPI*
Frequency	116	35	56	37	76	7	36
Percent	56	16.9	27.9	23.8	49	4.5	23.2
Valid Percent	56	16.9	27.9	23.8	49	4.5	23.2
	Laboratory Facilities				Water Facility		
	All Test Available	Limited Test Available	No Test Available	Outsourcing Test Facility	Purified Water	Ro Water	Tap Water
Frequency	67	70	19	51	100	79	28
Percent	32.3	33.8	9.1	24.6	48.3	38.2	13.5
Valid Percent	32.3	33.8	9.1	24.6	48.3	38.2	13.5

Source: Authors' calculation.

The research was conducted to illuminate the healthcare industry for both established players and new entrants. Patients have a resource for making decisions with these results. It may be used as a guideline while deciding on a hospital. In addition, researchers may apply the findings from this study in their work. This research was done in India as a case study to examine preferences and experiences in healthcare sector. It aims to give decision-makers in healthcare organisations useful information on how customers are responding to various aspects of the service they get. Understanding customer behaviour might be helpful for hospital management in making decisions about the future of the institution. It hopes to fill in some of the blanks in the literature between marketing, consumer behaviour, and healthcare, which might lead to useful information for future research. However, the general limitation is that we have only been able to study a small sample of responders because of time and money restrictions. The study's findings are consistent with those of another research, and it was carried out in Amritsar, Punjab (India). However, there is a possibility that our methods and findings will differ from those of other researchers. The use of a sampling technique based only on convenience is another restriction.

While the present study has made some strides in covering key areas of what consumers can expect from the healthcare sector, there is still a lot of room for exploration. As time goes on, more and more advances and recommendations are made available to enable healthcare managers set up a solid environment to deal with future pandemics. The verified framework that has been proposed allows for the study to be expanded to other parts of the country. Larger samples or alternative sampling strategies can be used in future studies. In addition, a disease-by-disease analysis may be performed to learn more about this topic.

10.8 CONCLUSION

A patient's disease is the primary driver of their lack of free will. When seeking a certain medical treatment, a patient often has several options to choose from. When deciding, he or she takes into account factors such as availability, price, and quality of service, as well as personal experience and recommendations from trusted sources (such as a doctor, friends, and family). A patient who has the financial means to do so should prioritise access to high-quality care. If all other factors remain constant, the patient will select the facility that is most conveniently located for them. However, the incorporation of human life and death only serves to further confuse matters. If patients believe that the treatments they are receiving are worthwhile, they may be ready to go further or pay more. It can be concluded that patients are willing to go further for a higher quality of care, including leaving their community to do so.

The most important factor in deciding where to receive treatment for an illness is how seriously ill one is. In cases of extreme urgency, patients will put aside concerns about how long it will take to receive care and the cost and quality of such care. Therefore, a knowledgeable provider (a doctor) plays a deciding role in a patient's decision. This doctor may practise in a variety of settings, including public, private, high-quality, low-quality, and nearby or far-removed facilities. Patients with more serious conditions opted for larger hospitals more frequently than those with less serious conditions. Patients should be able to voice their preferences to their chosen healthcare professionals. They need to be persuaded to prioritise health care quality when deciding where to go for treatment. Yet, when picking a hospital, people rarely look at hard data on the institution's actual performance. For patients to make an informed choice about hospital care, they need access to accurate and relevant information on service quality. Data should be used to back up patient preferences. All the research objectives were hence achieved, and all the hypotheses were very well justified in the above research.

Patient expectations from hospitals after Covid-19 are shown by the current study. Patients have claimed that basic facilities, renown and quality, building and infrastructure, and simplicity and cost are the most significant in their choice of hospitals. Patients have also thought about things like using their drugs, receiving helpful ideas and suggestions, and having access to clinical assistance. Patients also place a premium on hospitals' commitment to confidentiality during procedures, open communication with patients, and a comprehensive set of services. Healthcare organisations might utilise these elements when developing or reworking their commercial strategy. It might also be useful for hospital management to learn more about the factors that influence patients' decisions to visit a given facility.

Patient happiness with their hospital experience can be influenced by several things. These aspects of the service include the following: basic infrastructure and amenities, fulfilment of clinical requirements, separate arrangements for Covid patients, preventive measures to lower the spread of the disease, necessary measures to deal with crises similar to Covid, cost-effectiveness, ease of access, registration, and other administrative procedures. Doctors and other medical professionals may improve service quality and patient experiences by concentrating on these areas. Patients' satisfaction with hospitals might rise if they receive better care.

Patients' considerations while choosing a medical centre must be understood. The quality of a hospital's services, including its basic conveniences, registration and administrative processes, nursing and staff care, specialty doctors, complaint resolution process, and clinical assistance, can have a significant impact on a patient's overall hospital experience. The organisations might look at these elements as chances to deliver healthcare that goes above and beyond what the patients anticipate. By developing and providing services of the highest possible quality, healthcare organisations may better meet the needs of their clients.

REFERENCES

Barnartt, Sharon N. "A review of medical sociology textbooks." *Teaching Sociology* 18, no. 3 (1990): 372–376.

Boscarino, Joseph, and Steven R. Stelber. "Hospital shopping and consumer choice." *Journal of Health Care Marketing*, no. 2 (1982): 2–10.

Buzgan, Turan, and O. Güner. "The effectiveness of the world health organization in pandemics and the future of the post-pandemic era." *Reflections on the pandemic in the Future of the World* (2020): 95–114.

Cutler, David M., Mark McClellan, and Joseph P. Newhouse. "How does managed care do it?" *The Rand Journal of Economics* (2000): 526–548.

Estrela, Vania V., Asif Ali Laghari, Ricardo T. Lopes, Abdullah Ayub Khan, Shoulin Yin, and Anand Deshpande. "Introduction to intelligent healthcare in a post-pandemic world." In *Intelligent healthcare systems*, pp. 3–23. CRC Press, 2023.

Flexner, WA, EN, Berkowitz, and M, Brown. "Strategic planning in health care management." *In Strategic planning in health care management* (1981): 408–408.

Gaynor, Martin, James B. Rebitzer, and Lowell J. Taylor. "Physician incentives in health maintenance organizations." *Journal of Political Economy* 112, no. 4 (2004): 915–931.

Ghosh, Manimay. "An empirical study on hospital selection in India." *International Journal of Health Care Quality Assurance* 28, no. 1 (2015): 27–39.

Glied, Sherry. "Managed care." In *Handbook of health economics*, 2000, vol. 1, pp. 707–753. Elsevier.

Mohammad Mosadeghrad, Ali. "Patient choice of a hospital: Implications for health policy and management." *International Journal of Health Care Quality Assurance* 27, no. 2 (2014): 152–164.

Shah, Akeel A., R. S. R. W. Tangirala, R. G. A. W. R. Singh, R. G. A. Wills, and F. C. Walsh. "A dynamic unit cell model for the all-vanadium flow battery." *Journal of the Electrochemical Society* 158, no. 6 (2011): pp-A671.

Singh, Aarti, and Ratri Parida. "Decision-making models for healthcare supply chain disruptions: review and insights for post-pandemic era." *International Journal of Global Business and Competitiveness* 17, no. 2 (2022): 130–141.

Verma, Madhur, Kirtan Rana, Ankita Kankaria, and Ramnika Aggarwal. "Assessment of patient's satisfaction visiting a tertiary health care institute in north India." *Journal of Pharmacy & Bioallied Sciences* 12, no. 3 (2020): 252–261.

Whitehead, Stephen S., Joseph E. Blaney, Anna P. Durbin, and Brian R. Murphy. "Prospects for a dengue virus vaccine." *Nature Reviews Microbiology* 5, no. 7 (2007): 518–528.

Wu, Peilin, Zhenjing Li, Wei Guo, Li Wang, Xiangxiang Chang, Yanqun Zhang, Li Wang, Lidan Wang, and Qunying Liu. "Optimizing the practice environment for medical staff in the post-pandemic era: A discrete choice experiment." *Frontiers in Public Health* 10 (2022): 911868.

Yiu, Chris, Benedict Macon-Cooney, and Henry Fingerhut. "A research and policy agenda for the post-pandemic world." *Future Healthcare Journal* 8, no. 2 (2021): e198–e203

Chapter 11

Anthropogenic climate change and the escalating vulnerability of coastal communities

Rushali Mishra, Juhi Gupta and Akarsh Verma

11.1 INTRODUCTION

It is a well-known fact that climate change has caused some of the most devastating, disastrous conditions known to mankind. The sudden erratic changes were a result of the anthropogenic influence on the planet. Major anthropogenic influence on the environment and the climate systems began with the Industrial Revolution, about 150 years ago. Since then, significant quantities of pollutants have been released into the atmosphere. These include greenhouse gases (GHGs) such as carbon dioxide, methane, and nitrous oxide whose concentrations have increased by about 31%, 151%, and 17%, respectively, between 1750 and 2000 (Vijaya Venkata Raman et al., 2012) (Figure 11.1). This accelerated climate change at an unprecedented rate led to a worldwide threat. The increasing fatalities associated with climate change pose a huge number of alarming threats. It is expected that it may cause 250,000 deaths per year between 2030 and 2050. Rising temperatures worsen conditions and cause people to get sicker from side effects ranging from heatstroke to cardiovascular and kidney diseases (NRDC, 2021). Extreme conditions caused by climate change may lead to increased disasters, further worsening living conditions and deteriorating health conditions.

Major community displacements due to disasters are becoming more common, around 60 percent higher than four decades ago, where most seem to be weather and climate-related events (NRDC, 2021; Martinez et al., 2022). The degrading conditions, along with depleting fresh resources, reflect in the urban market as the economy takes a dip, leading to "double exposure" (Owen, 2020). Cost of living is seen to increase with it. Less developed countries are more affected than developed ones due to lack of access to resources and lack of protection from the climate (Tol, 2018). A greater understanding of climate change and its effects, especially over the last two decades, has led to increased resources going toward the cause of adaptation through development of new technologies, influencing decision-making, better infrastructure, increasing awareness, increasing access to resources and decreasing exposure to environmental risks (Owen, 2020). Climate change is a complex phenomenon but can be understood by the

DOI: 10.1201/9781003621102-11

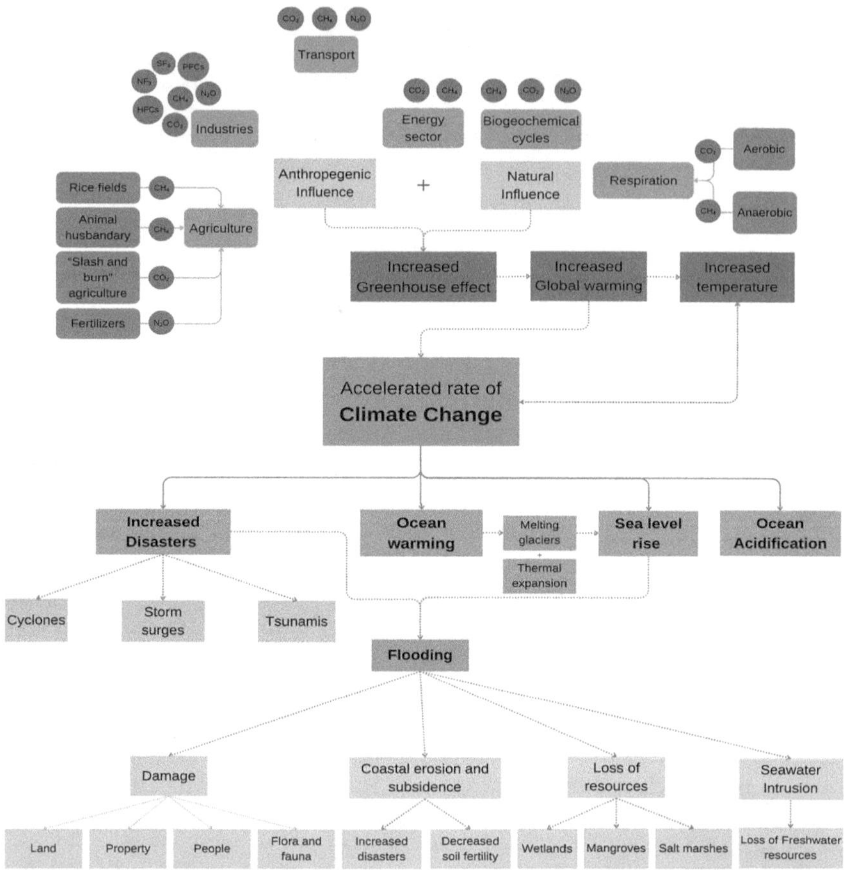

Figure 11.1 Sources of greenhouse gases and effects of climate change (Shakoor et al., 2020; Trivedi et al., 2020; NAEI, 2023; UN, 2024; EPA 2024).

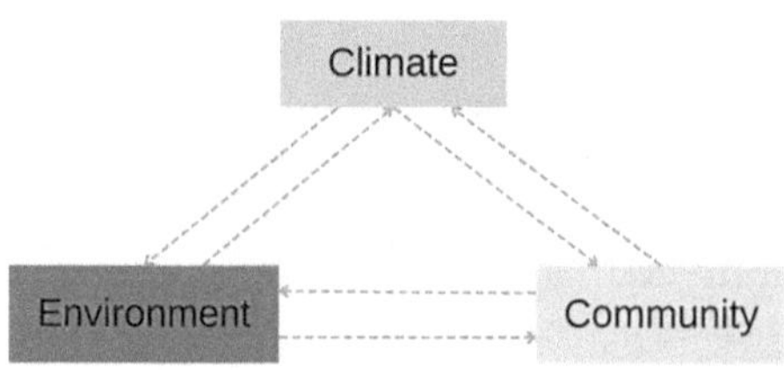

Figure 11.2 Interrelation between climate change, environment, and habitants of the area.

factors affecting it. Being a naturally occurring process, it has a dependence on the conditions of the environment, which is further dependent on the species present in it (Figure 11.2). Therefore, as the species change and start making permanent changes in their environment, it slowly starts to reflect in the local climate system of the area as well.

The effects of climate change are wide-ranging and difficult to control, but understanding the factors involved provides a clearer picture so that the consequences presented can be dealt with adequately.

11.2 MARINE ECOSYSTEM

The marine environment is one of the most significant ecosystems offering a variety of services to the coastal community. They support sectors such as fishing, industry, and transportation, attracting immigrants and tourists and leading to a high population density in such a resource-rich area. The vulnerability of coastal areas is due to their geographical location and other characteristics, such as their high population density. The coastal-marine ecosystem is a fragile, dynamic ecosystem with subtle changes resulting in changes in major characteristics of the ocean. Greater use and depletion of resources, along with increased pollutants, come with a greater population density, with the demand even higher in urban scenarios. These may eventually lead to losses in major problems and losses in important ecosystem services, leading to greater competition and social conflicts over resources (Barragán and De Andrés, 2015). Low-latitude developing countries are expected to be at the forefront of experiencing the catastrophic effects of climate change (Hasan and Nursey-Bray, 2018). Aquatic life is experiencing changes in water composition due to changes in temperature and pH because of climate change. This disrupts migratory patterns (example: salmon) and the reproductive rate of eggs hatching, among other things. Combined with harmful algal blooms boosted by increased nutrient levels released from rocks and soil present in warm waters depletes the oxygen levels of the ocean leading to disruptions throughout the marine ecosystem (Rodgers, 2021). The declining number of fish and other aquatic life affects the fishing sector, a major source of employment and nutrition in the coastal areas. Tourism, another source of income in the coastal communities, also reduces with increasing risks and degrading conditions. The overall economy of the region deteriorates, leading to a more vulnerable society.

The coastal regions are rich in natural resources that are utilized by the communities present in them. The marine environment provides food and nutrition, employment, and other resources due to which the coastal communities have grown very reliant on them. It forms an essential part of their lives getting integrated into their culture as well which is protected and preserved for generations. Various services fulfilled include nutritional, economic, cultural as well as coastal protection services (Selig, et al., 2019; (NOAA, 2024):

1. Nutritional: Critical source of lipids and micronutrients. Fish and other marine organisms form a large part of the diet as they are one of the more readily available food options and feasible to acquire. They form a major part of the food intake due to food preference

(e.g. Japan) or due to lack of other options and malnutrition (e.g. some West African countries).

2. Economical: Support more than 260 million livelihoods; generate large export revenues for developing countries (80 billion USD) (FAO, 2016). Fishing culture forms a big part of their lives, and they utilize it not only for replenishment but also for selling. Fishing is a major job sector in rural communities living near water.
3. Cultural: Generations living near the coasts led them to depend on them in more than a few ways. Daily practices integrate into the culture of the residing communities. Spiritual and religious connections may form over time.
4. Coastal protection: The oceans and seas are dynamic and dangerous in nature. Mangroves, coral reefs, etc. act as a buffer between the violent sea tides and waves and the coasts by absorbing the impact so that the coastal region doesn't have to, leading to reduced disasters and impacts. Corals reduce wave height up to 70 percent.

Countries present in the Pacific and the Indian ocean, countries outlining the western coasts of Africa, and a few Southeastern Asian countries (e.g., Indonesia and Philippines) displayed highest nutritional and economic dependence (Selig et al., 2019). About 775 million people were estimated to be relatively highly dependent on marine ecosystems, of which 525 million people's dependence was nutritional. This highlights the importance of fisheries in today's scenario and its relationship with food security. Sustainable resource management should be practiced (Selig et al., 2019) to lead to a healthy economy and healthy, growing communities. Climate change has displayed various effects on the components of the coastal communities. Effects such as ocean warming and ocean acidification have been seen to severely impact the aquaculture of fisheries and shellfish in some coastal areas (IPCC, 2022), which further impacts the food and job security of the region.

A limit has been set on global warming worldwide of 1.5°C limit by 2030, after which serious consequences may follow. Freshwater resources acquired by glaciers and snow continue to deplete along with inland freshwater resources which are getting contaminated with seawater due to flooding. Upon reaching hard adaptation limits for warm water coral reefs, coastal wetlands, etc., ecosystem-based adaptation will lose its effectiveness and no longer be viable as an option (IPCC, 2022).

11.3 ADAPTATION TO COASTAL AREAS

Adaptation can be defined as an adjustment in natural or human systems due to present or future climate changes to moderate the harm caused by the effects or even gain benefits from them (EEA, 2012) (Figure 11.3).

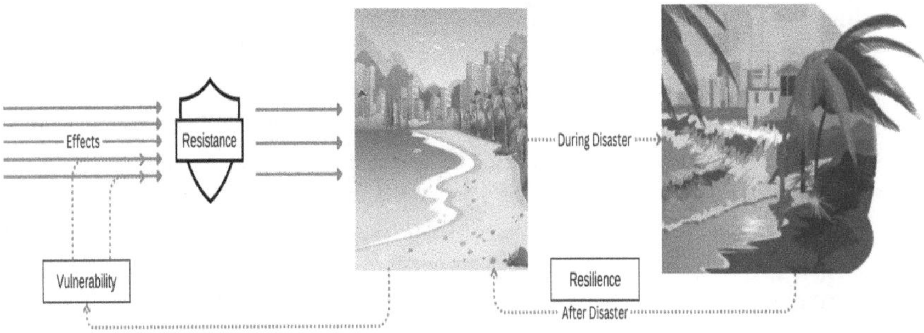

Figure 11.3 Different aspects of adaptability: resistance, resilience, reduced vulnerability.

Adaptation strategies have to fulfill one or more of the following goals:

- Improving resistance: Forestall the impacts and protect valuable resources from it (Millar et al., 2007).
- Increasing resilience: Change with oncoming stress but retain the ability to revert back to their original state with continued functioning (Lawler, 2009; Millar et al., 2007).
- Improving response: Gradually transition to accommodate the changes occurring in the climate to avoid rapid changes, leading to loss of ecosystem functioning, species, and more (Lawler, 2009; Millar et al., 2007).
- Decrease vulnerability: Reducing or removing possible factors of vulnerabilities will control the impact faced by the region.
- Increase overall adaptive capacity.

Different regions have different conditions they are exposed due to which they may require adaptation processes adjusted to their needs. Coastal areas of various countries around the world have adopted different adaptation techniques to battle climate change.

11.4 CASE STUDIES OF AFFECTED COASTAL ZONES AND ASSOCIATED DISASTERS

11.4.1 Bangladesh

Bangladesh is one of the countries that is seen to be more vulnerable to the effects of climate change and is termed to be at "extreme risk". Due to its geographical location, it experiences increased frequency and intensity of disasters, receiving about 40 percent of the total impacts of global storm surges. The resulting flooding causes coastal erosion, coastal inundation, and rising levels of salinity, compromising crops, biodiversity, and safe drinking water which in turn affects the health, economy, and the flora and fauna present in the region (Hasan and Nursey-Bray, 2018).

Fisheries rank as the second most important agricultural sector in Bangladesh, providing full-time employment to about 1.2 million fishers, consisting of the fourth largest inland fisheries production in Asia, following China, Myanmar, and India. It also forms the main animal protein for 60% of the population, supplying about 160 million residents of the country. In addition to providing income, food security, nutritional and economic benefits, fishing also finds its place in Bangladesh's culture and tradition (Hasan and Nursey-Bray, 2018; Hossain et al., 2018). Severe levels of poverty, dependence on agriculture, and low adaptive capacity of its government make the country more susceptible to the effects of climate change. Their location and socioeconomic status determine the level of vulnerability where a few classes live in cement brick houses while the others live in temporary houses made of bamboo, mud, tin, and even plastic which hold no hope for surviving the fatal disasters faced in the region (Hasan and Nursey-Bray, 2018).

Cyclone Alia

In 2009, Cyclone Alia, with an average speed of 95 kmph, hit Bangladesh, leaving 190 dead, 7103 injured, overall affecting about 3.9 million people. Total damage to assets was estimated to be around 270 million dollars. Almost all plants seemed to be destroyed, the land salinized and its fertility dwindling, making it unusable for agriculture and fish farming for years to come. Still, its consequences can be seen reflecting on the abundance and productivity of the existing biodiversity and livelihood of aquatic ecosystems of the region (Hossain et al., 2018).

Since 1970, 36 cyclones have hit Bangladesh, causing over 450,000 deaths and immense economic losses (Hossain et al., 2018).

Different adaptation techniques adopted include:

1. Migration: Migration for a few days till conditions improve, or permanent migration of better opportunities and income is seen. Leaving their profession behind proves costly as they face underemployment and face even harsher, unstable conditions in urban cities. Shrimp farmers left in great numbers due to increased input prices and lease (Hossain et al., 2018; Hasan and and Nursey-Bray, 2018).
2. Government incentives: Selective and limited food incentives in hilsa districts (40 kg rice a month for 4 months) proved insufficient for the fisher families. Defying fishing regulations and payment of penalties therefore follow.
3. Alternative income source/Livelihood diversification: Promotion of non-fishing-related activities to provide an alternative source of income during seasonal and inter annual-variation in fish availability. Example: farming (Hossain et al., 2018).
4. Decreasing livelihood vulnerability: Vulnerabilities due to gender imbalance, salinization of agricultural land, decreased access to

cooking fuels, and sufficient clean drinking water exist. Lack of credit support from the government holds back women from taking up alternate professions (Hossain et al., 2018).

5. Sustainable fisheries management: Reducing social stresses such as poverty, inequality,, and environmental stresses (pollution, overfishing, and habitat destruction) decreases vulnerability and increases resilience while also increasing global production of fisheries by around 10% (Hossain et al., 2018;).
6. Planned adaptation: Empowering the fishing community by planning according to their needs, abilities, and interests enhances cooperation as well due to the feeling of a strong sense of belonging and understanding. Strengthened social relationships lead to better adaptation in the face of adversity (Hossain et al., 2018).
7. Spatial management/protection for conservation: Bangladesh has more than 500 established fish sanctuaries such as Saint Martin Island and the Sundarbans (mangrove forest) and a marine reserve of area 698 km^2 in the Bay of Bengal for the protection and preservation of various breeding grounds. Marine spatial planning analyzes and allocates the spatial and temporal distribution of human activities in marine ecosystems (Hossain et al., 2018).
8. Ecological restoration: Rehabilitating ecosystems and habitats by increasing diversity and connectivity leads to a more resilient, efficient ecosystem (Hossain et al., 2018).
9. Community-based adaptation (CBA): The community comes together pre-disaster and post-disaster for preparations and relief (Ahmed et al., 2017).

11.4.2 Philippines

Philippines, a country made up of more than 7000 islands, is heavily dependent on fishing and aquaculture. The sector seems to be highly unstable and unable to support the large population of inhabitants of the country which are more than 100 million in number. A study found 30 (of 44) Philippine towns' fisheries present in municipal waters to be "bound to collapse" due to the pressure they were under (Muallil et al., 2014). This tropical archipelago faces direct as well as indirect effects of climate change. The physical effects include temperature, winds, and vertical mixing of the marine waters, whereas the chemical effects include their levels of salinity, oxygen, and the pH of the water (Jacinto et al., 2015). The National Framework Strategy on Climate Change was formed to counteract climate change and its devastating effects on the Filipino community. The importance of vulnerability assessments (VAs) was highlighted (Jacinto et al., 2015), especially for the fisheries and aquaculture communities, to understand the ecosystem, the interactions within, and the threats posed to it, thus forming the base on which adaptation options can be built upon (Mamauag et al., 2013).

Adaptation strategies adopted include:

1. Gear modification/replacement for effectiveness, adaptation: Resources such as a sufficient budget and gear and the associated knowledge are rewarded with higher catch rates, lesser fuel, and time spent on fishing activities (Jacinto et al., 2015).
2. Precautionary measures practiced by the fishers before, during, and after disasters (Jacinto et al., 2015).
3. A level of awareness of climate change helps fishers plan their activities around them accordingly. Seminars, workshops, and training can help educate on climate change and the possible impacts, and more importantly ways to deal with them through adaptation, risk reduction, and diversification of livelihood (Jacinto et al., 2015).
4. Literacy and alternative skills help the fishers find other sources of income, either in addition to fishing or as a replacement for it. It increases the possible options they can look to, especially when disaster strikes and job opportunities lessen (Jacinto et al., 2015).
5. Strong community support system: Community relations and support during challenging times can help boost morale and help get through tough times (Jacinto et al., 2015).
6. Access to information: To serve as some form of an early warning system, source of information, knowledge through TV, radio, newspaper, internet, school, etc. to prepare accordingly (Jacinto et al., 2015).
7. Annual income from fishing: Higher levels of average income indicate high dependence on fishing activities, reflecting the community's poor ability to adapt when their primary source of income is endangered (Jacinto et al., 2015).

11.4.3 Pakistan

Mangrove forests are known to be one of the most indispensable resources of the coastal areas: from supporting a considerable number of species with varying degrees of ecological and economic viability. Indus Delta of Pakistan, which hosts about 97% of total mangrove forests (Abbas et al., 2013), provides not only the livelihoods of many but provides a substantial contribution to the nation's economy through the export of marine fish. The threat of degradation and extinction is presented to them due to various climatic (changes in temperature, precipitation, sea surface temperature, sea level rise, extreme weather events) as well as non-climatic factors such as low socioeconomic conditions (low education, lack of healthcare facilities, low income diversification, high dependency ratio, absence of water and sanitation conditions) which may eventually lead to exploitation and unsustainable use of resources as desperate measures in uncertain times (Salik et al., 2015). The National Climate Change Policy (NCCP), 2013, was put forward by the Government of Pakistan stating all climate change matters in the country.

Different adaptation practices undertaken include:

1. Provision of safe water drinking facilities: The groundwater in the delta is saline, making it unsuitable for drinking and causing residents to drink from the untreated irrigation canals contaminated with industrial and agricultural effluents, posing high risks to health. This water is not always available, mostly as a result of season closure, which is at least 5 months in a year, and agricultural use. This also causes temporary as well as permanent out-migration (Salik et al., 2015).
2. Ensuring freshwater flows: Diminishing freshwater flows have contributed to the degradation of the coastal ecosystem and its resources, resulting in less livelihood options and further worsening the issues of poverty and migration (Salik et al., 2015).
3. Adaptation strategies for high-risk areas: The small and scattered villages around the creeks face a high risk of storms and floods and could benefit from specific and clearly defined adaptation and protection strategies to better deal with the oncoming disasters (Salik et al., 2015).
4. Improving access to education: Better educational opportunities not only enhance awareness regarding risks and vulnerabilities but tackle poverty by increasing and improving livelihood options (Salik et al., 2015).
5. Development and implementation of climate change adaptations and activities (Salik et al., 2015).

11.4.4 Malaysia

Fishing has a as a significant role to play to in employment and economic growth in Malaysia, in especially rural areas (Yaakob and Quah, 2005). Extreme waves caused by the rising sea level have been seen to cause coastal erosion. Mangrove forests are destroyed due to erosion at a rate of 0.8% per year, while 29.1% of coastal areas are experiencing critical erosion. In a 14-year period (1986–2000), the area experiencing critical erosion increased from 156 km to 233 km (Shaffril et al., 2017). Ocean warming affects spawning rates and migration patterns as well as encourages the growth of harmful algal blooms affecting fish species such as sardines. Changes in wind and precipitation patterns have affected fishing rates, reducing up to 90% of operating months of fishermen in a month during the northeast monsoon. A study showed a 32% reduction in catches due to climate change (Shaffril et al., 2017).

Adaptation practices include:

1. Reducing vulnerabilities of fishermen to extreme weather and financial issues: Earnings of almost 1/3rd Malaysian fishermen lies around 500RM due to which the National Fishermen Association (PNK) has introduced an insurance protection scheme worth about 100 RM per year (Shaffril et al., 2017).

2. Stronger community: Strengthening social relationships in communities shows great promise, especially during pre-disaster and post-disaster preparation (Shaffril et al., 2017).
3. Generating awareness: With the help of governmental and non-governmental stakeholders such as the Community-Based Disaster Risks Management (CBDRM) program by Mercy Malaysia, sufficient awareness is generated. It focuses on increasing the involvement of the local government and the community to identify and analyze potential risks in their environment and address them. This program exists in China, the Maldives, Myanmar, and Malaysia (Shaffril et al., 2017).
4. Alternative skills: Learning alternative skills that can contribute to one's livelihood can contribute to the betterment and resilience of the communities. Force of Nature Aide Foundation is a non-profit agency which looks to empower and rebuild post-disaster communities. Their collaborations with the World Fish Centre, Aceh, has proven the success of their strategies (Shaffril et al., 2017).
5. Involvement of community: Involvement in adaptation planning strategies ensure that the plans align with the community's needs and wants. It also checks for feasibility within the local context. Examples include the PNK's annual meeting for registered fishermen who attend and contribute suggestions and views to the relevant agencies (Shaffril et al., 2017).
6. Improving small-scale fishermen's access to credit to help in adaptation and recovery: Agencies such as Agro Bank, the LKIM, AIM, and PNK offer credit with low to no interest rates (Shaffril et al., 2017).

Derived adaptation strategies:

1. Restructuring of coastal cities: Increasing the resilience and resistance of the coastal areas should start with restructuring of the cities, given their dense population and increasing vulnerability.
2. Creation of zones: Creation of zones along the shores should also be made a priority, in addition to restoration of natural ecosystems such as mangroves and wetlands to help absorb some of the incoming impact.
3. Early warning systems and flood proofing
4. Urban adaptation techniques include disaster risk management, social safety nets, and green/blue infrastructure like levees to reduce inland flooding.
5. Ecosystem-based adaptation techniques are also useful in increasing the natural resilience provided by ecosystems and biodiversity of the area which makes use of mangroves, coastal wetlands, etc. These approaches include urban greening, restoration of wetlands, and upstream forest ecosystems, among a few others. They help reduce flood risks, urban heating, and many other benefits. Minimizing or removing anthropogenic stresses, reducing fragmentation, increasing extent of habitats, and protecting small-scale refugia play an important role,

6. Land use planning such as no build zones or upstream forest management can help decrease flood risk.
7. Proper adaptation techniques can help improve agricultural productivity, innovation, health and well-being, food security, livelihood, and biodiversity conservation while reducing the risks and damages associated with the various effects of climate change on coastal communities.
8. Diversification of livelihood options by engaging in alternative skills or similar skills such as inland fishing and marine aquaculture or picking up a second job can help increase job security and options.

11.5 CONCLUSION

Climate change cannot be stopped, given it's a natural phenomenon occurring worldwide and affecting diverse parts of our community. Adaptation is a process of transition to safer, more protected areas that can withstand the harsh effects of climate change to an extent. Reduction in the intensity of impact can help in the process of recovery and planning for the next calamity that occurs. It's a continuous cycle with no visible end point (Owen, 2020). Three main approaches of adaptation techniques that can be adopted include accommodation, protection, and retreat (Gargiulo et al., 2020), which have been observed and reported in many coastal areas. Short-term practices tend to be more popular due to the immediate effects seen, but long-term practices look to preserve societies from the effects to an extent, growing stronger over time. Adopting and contributing to adaptation practices ensures safety in the long run.

Adaptive strategies are rarely implemented, even though planned and proposed, due to uncertainty, lack of agreement, and attribution. Other factors that play a role are insufficient resources, prohibitive policies, different, often competing or conflicting, priorities for action, and uncertain future changes (Owen, 2020). Taking the regional politics, funding, cultural values, and other characteristics makes for more well-rounded adaptive strategies than general ones. However, general practices can act as a good foundation to adjust and build upon accordingly (Owen, 2020).

Routine evaluation and analysis can determine the suitability of adaptive processes to check for the degree of uncertainty, disagreement, and attribution. Evaluation and improvement over time could result in suitable, successful adaptation (Owen, 2020). Current scenarios can be dealt with knowledge and data gathered over the years. Uncertainty of the changes occurring in the future create a complex issue to be resolved, but strategies that are robust, safe, and overall cost-effective can be implemented to be mended over time accordingly (Spalding et al., 2014). Adaptation will not remain a viable option for long. As conditions continue to deteriorate, their effectiveness will reduce significantly (IPCC, 2022). Therefore, adaptation practices accompanied by complimenting mitigation techniques increase the possibility of a better future.

ACKNOWLEDGMENT

The authors are grateful to Amity Institute of Environmental Sciences, Amity University, Noida, for their academic support.

CONFLICT OF INTEREST STATEMENT

The authors declare that there is no conflict of interest.

REFERENCES

Abbas, S., Qamer, F.M., Ali, G., Tripathi, N.K., Shehzad, K., Saleem, R. and Gilani, H., 2013. An assessment of status and distribution of mangrove forest cover in Pakistan. *Journal of Biodiversity and Environmental Sciences*, 3(6), pp.64–78.

Ahmed, A.U., Mondal, P. and Islam, M., 2017. *Community-based Adaptation: An Analysis of Best Practices in the South-western Region of Bangladesh*, Dhaka, CARE Bangladesh, pp. 70.

Barragán, J.M. and De Andrés, M., 2015. Analysis and trends of the world's coastal cities and agglomerations. *Ocean and Coastal Management*, 114, pp.11–20.

Environmental Protection Agency (EPA). 2024. Overview of greenhouse gases, https://www.epa.gov/ghgemissions/overview-greenhouse-gases#N2O-references. Accessed 20 May 2024.

European Environment Agency (EEA). Urban adaptation to climate Change in Europe. Challenges and opportunities for cities together with supportive national and European policies. 2012.

Food and Agriculture Organization (FAO). 2016. *The State of World Fisheries and Aquaculture 2016*. Contributing to food security and nutrition for all. Rome. 200 pp.

Gargiulo, C., Battarra, R. and Tremiterra, M.R., 2020. Coastal areas and climate change: A decision support tool for implementing adaptation measures. *Land Use Policy*, 91, p.104413.

Hasan, Z. and Nursey-Bray, M., 2018. Artisan fishers' perception of climate change and disasters in coastal Bangladesh. *Journal of Environmental Planning and Management*, 61(7), pp.1204–1223.

Hossain, M.A., Ahmed, M., Ojea, E. and Fernandes, J.A., 2018. Impacts and responses to environmental change in coastal livelihoods of south-west Bangladesh. *Science of the Total Environment*, 637, pp.954–970.

Irwin, W.P. and Barnes, I., 1980. Tectonic relations of carbon dioxide discharges and earthquakes. *Journal of Geophysical Research - Solid Earth*, 85(B6), pp.3115–3121.

Jacinto, M.R., Al Jayson, G.S., Von Yip, G. and Santos, M.D., 2015. Development and application of the fisheries vulnerability assessment tool (fish vool) to tuna and sardine sectors in the Philippines. *Fisheries Research*, 161, pp.174–181.

Lawler, J.J., 2009. Climate change adaptation strategies for resource management and conservation planning. *Annals of the New York Academy of Sciences*, 1162(1), pp.79–98.

Mamauag, S.S., Aliño, P.M., Martinez, R.J.S., Muallil, R.N., Doctor, M.V.A., Dizon, E.C., Geronimo, R.C., Panga, F.M. and Cabral, R.B., 2013. A framework for vulnerability assessment of coastal fisheries ecosystems to climate change—Tool for understanding resilience of fisheries (VA–TURF). *Fisheries Research*, 147, pp.381–393.

Martinez, G.S., Kendrovski, V., Salazar, M.A., de'Donato, F. and Boeckmann, M., 2022. Heat-health action planning in the WHO European region: Status and policy implications. *Environmental Research*, 214, p.113709.

Millar, C.I., Stephenson, N.L. and Stephens, S.L., 2007. Climate change and forests of the future: managing in the face of uncertainty. *Ecological Applications*, 17(8), pp.2145–2151.

Muallil, R.N., Mamauag, S.S., Cabral, R.B., Celeste-Dizon, E.O. and Aliño, P.M., 2014. Status, trends and challenges in the sustainability of small-scale fisheries in the Philippines: Insights from FISHDA (Fishing Industries' Support in Handling Decisions Application) model. *Marine Policy*, 44, pp.212–221.

National Atmospheric Emissions Inventory (NAEI). Greenhouse gases, https://naei.energysecurity.gov.uk/greenhouse-gases. 4 October, 2023.

National Oceanic and Atmospheric Administration (NOAA). What threats do coastal communities face?, https://oceanservice.noaa.gov/facts/coastalthreat.html. 16 June 2024.

Owen, G., 2020. What makes climate change adaptation effective? A systematic review of the literature. *Global Environmental Change*, 62, p.102071.

Rodgers, E.M., 2021. Adding climate change to the mix: Responses of aquatic ectotherms to the combined effects of eutrophication and warming. *Biology Letters*, 17(10), p.20210442.

Salik, K.M., Jahangir, S. and ul Hasson, S., 2015. Climate change vulnerability and adaptation options for the coastal communities of Pakistan. *Ocean and Coastal Management*, 112, pp.61–73.

Selig, E.R., Hole, D.G., Allison, E.H., Arkema, K.K., McKinnon, M.C., Chu, J., de Sherbinin, A., Fisher, B., Glew, L., Holland, M.B. and Ingram, J.C., 2019. Mapping global human dependence on marine ecosystems. *Conservation Letters*, 12(2), p.e12617.

Shaffril, H.A.M., Samah, A.A. and D'Silva, J.L., 2017. Adapting towards climate change impacts: Strategies for small-scale fishermen in Malaysia. *Marine Policy*, 81, pp.196–201.

Shakoor, A., Ashraf, F., Shakoor, S., Mustafa, A., Rehman, A. and Altaf, M.M., 2020. Biogeochemical transformation of greenhouse gas emissions from terrestrial to atmosphleric environment and potential feedback to climate forcing. *Environmental Science and Pollution Research*, 27, pp.38513–38536.

Spalding, M.D., McIvor, A.L., Beck, M.W., Koch, E.W., Möller, I., Reed, D.J., Rubinoff, P., Spencer, T., Tolhurst, T.J., Wamsley, T.V. and Van Wesenbeeck, B.K., 2014. Coastal ecosystems: a critical element of risk reduction. *Conservation Letters*, 7(3), pp.293–301.

Tol, R.S., 2018. The economic impacts of climate change. *Review of environmental economics and policy*.

Trivedi, S., Prasad, R., Mishra, A., Kalam, A. and Yadav, P., 2020. Current scenario of CNG vehicular pollution and their possible abatement technologies: An overview. *Environmental Science and Pollution Research*, 27, pp.39977–40000.

United Nations (UN). 2024. What is climate change, https://www.un.org/en/climatechange/what-is-climate-change. Accessed 20 May 2024.

Vijaya Venkata Raman, S., Iniyan, S. and Goic, R., 2012. A review of climate change, mitigation and adaptation. *Renewable and Sustainable Energy Reviews*, 16(1), pp.878–897.

Yaakob, O. and Quah, P.C., 2005. Weather downtime and its effect on fishing operation in Peninsular Malaysia. *Jurnal Teknologi*, pp.13–26. https://doi.org/10.11113/jt.v42.730

Index

Kolibri 360 GmbH
Ferdinand-Jühlke-Straße 7
99095 Erfurt, Deutschland
produktsicherheit@kolibri360.de